# BASIC WELDING

FOR

MOTIVE POWER TECHNICIANS

Sean Bennett

Centennial College Press

Copyright © 2003 by Sean Bennett

All rights reserved. No part of this book may be reproduced in any form without written permission of the copyright owners. All images in this book have been reproduced with the knowledge and prior consent of the companies concerned and no responsibility is accepted by producer, publisher, or printer for any infringement of copyright or otherwise, arising from the contents of this publication. Every effort has been made to ensure that credits accurately comply with information supplied.

Cover Design:    Larry Farr, Centennial College
Ross Maddever

Photo Cover:    Courtesy of Lincoln Electric Company of Canada

Published by:    Centennial College Press
P.O. Box 631
Station A
Toronto, ON
M1K 5E9

Website: www.thecentre.centennialcollege.ca
(services/resources)
Sean Bennett: www.seanbennett.org

ISBN: 0-919852-43-2

Printed and bound in Canada

# Contents

| | | |
|---|---|---|
| Introduction | ................................................ | v |
| Chapter 1 | Welding Safety .................................. | 1 |
| | ➢ Personal Safety / 2 | |
| | ➢ Shop Safety / 4 | |
| | ➢ Equipment Safety / 5 | |
| | ➢ Welding Helmets / 7 | |
| | ➢ Master Chart of Welding and Allied Processes / 8 | |
| | ➢ UV Radiation / 9 | |
| | ➢ Hand Tools / 10 | |
| | ➢ Fire Categories / 11 | |
| | ➢ Types of Fire Extinguishers /12 | |
| | ➢ Suppressing Fires / 13 | |
| Chapter 2 | Oxy-acetylene Processes ........................ | 15 |
| | ➢ Acetylene / 15 | |
| | ➢ Oxygen / 17 | |
| | ➢ Regulators and Gauges / 19 | |
| | ➢ Hoses and Fittings / 19 | |
| | ➢ Backfire / 19 | |
| | ➢ Flashback / 20 | |
| | ➢ Torches and Tips / 20 | |
| | ➢ Eye Protection / 21 | |
| | ➢ Oxy-acetylene Precautions / 21 | |
| | ➢ Adjustment of the Oxy-acetylene Flame / 21 | |
| | ➢ Oxy-acetylene Welding Practice / 23 | |
| Chapter 3 | Gas Cutting of Steel ............................. | 25 |
| | ➢ Cutting Torch / 25 | |
| | ➢ Oxy-acetylene Flame Adjustment / 28 | |
| Chapter 4 | Arc Welding ..................................... | 29 |
| | ➢ Arc Welding Principles / 29 | |
| | ➢ Polarity / 29 | |
| | ➢ Alternating Current Welding / 32 | |

- Arc Welding Equipment / 32
- Electrodes / 35
- Function of Ingredients in Electrode Coatings / 39
- Electrode Selection Guide / 41
- Arc Welding Practice / 44
- Elements of Welding Practice / 48

| Chapter 5 | Metal Inert Gas (MIG) Welding | 51 |
|---|---|---|

- MIG Process / 51
- Shielding Gases / 54
- Welding Equipment / 55
- MIG Welding Techniques / 59

| Chapter 6 | Tungsten Inert Gas (TIG) Welding | 67 |
|---|---|---|

- Introduction / 67
- TIG Principle of Operation / 68
- TIG Equipment / 69
- TIG Welding Practice / 75
- TIG Technique / 76

| Chapter 7 | Metal Basics | 77 |
|---|---|---|

- Metal Properties and Identification / 77
- Alloys Used to Improve the Metallurgical Characteristics of Steel / 79
- SAE Classifications of Steels / 81
- Other Metals and Metal Alloys / 81

| Chapter 8 | Welding Techniques | 85 |
|---|---|---|

- Introduction / 85
- Oxy-acetylene Flame Cutting Exercise / 86
- General Chassis Repairs / 88
- Welding Vehicle Fuel Tanks / 89
- Welding a 6" Pipe in Position / 93
- Lengthening a Truck Frame Rail / 95

| Glossary | | 99 |
|---|---|---|

- Techniques / 99
- Weld Faults / 100
- General Welding Terminology / 101

# Introduction

This text is intended to provide motive power technicians with an introduction to welding technology. Welding is one of the many skills that are part of the practice of being an automotive, truck, trailer or heavy equipment technician. The extent to which a technician is required to weld depends to on the specific branch of motive power technology that technician is employed in. For instance, the average truck trailer technician will generally be required to perform more welding and to a higher standard than the typical automotive service technician. In large service facilities especially those specializing in heavy duty on-highway and off-highway equipment, employment of a specialty welder is often justified. This becomes even more necessary, when such items as hardened steel frame rails and high-tech alloy metals have to be welded.

**Objectives**
The objective of this book is to provide the basics of welding technology required by the auto, truck or heavy equipment technician without giving the subject matter the kind of in-depth coverage required for professional specialty welding. It is structured to address the knowledge level required in apprenticeship programs in which welding is one of a multitude of prerequisite skills. The text begins with some basic welding safety practices before describing welding equipment, the theory of fusion welding, some basic metallurgy, and welding practice.

# WELDING SAFETY 1

The practice of welding should always be safe. No element of risk can be considered acceptable. Welding safety is simply a matter of never compromising a few safety rules and checks, and above all, recognizing the potential of careless practice and malfunctioning equipment to cause serious and even fatal injuries. The welder first must observe personal safety. Next, the welder must recognize that unsafe practice can injure anyone in the vicinity of the welding area. Also important is the maintenance of welding equipment which can be costly if neglected, and maintaining a safe shop environment. Safety issues can be generally divided into the following categories:

### ➢ Personal Safety
Personal safety covers anything that affects the safety of the operator of the welding equipment.

### ➢ Shop Safety
This important area covers the safety of any person in the vicinity of the welding being performed.

### ➢ Equipment Safety
Malfunctions in equipment can cause accidents and long term health risks.

# PERSONAL SAFETY

Without doubt, this is the most important safety consideration. Statistically, most welding incident injuries affect the operator of the welding equipment. Personal safety in welding practice is largely a matter of common sense but it never hurts to underline the basics when the consequences can be serious injury or death. Here are some guidelines:

SUITABLE CLOTHING

➢ Wear clothing suitable for welding. Clothing that is highly combustible must be avoided. This means avoiding man-made fibers. Man-made fibers can be highly combustible and many will melt to a plasticized liquid that fuses to skin, causing severe burns. Wearing natural fibers such as cotton and wool, with additional protection such as leather welding jackets, apron and spats provides maximum protection. For obvious reasons, positional welding requires a greater degree of leather clothing protection.

➢ Wear leather gloves or gauntlets designed for welders. Ensure the leather is soft and pliable and not oil saturated.

➢ Wear safety boots rather than shoes. Never wear socks made of man-made fiber and do not tuck coveralls into socks above the boot. Ensure that the boot is laced tight to the ankle. Many severe burns are caused by molten lumps of metal falling into an open shoe or boot.

➢ Cover all exposed skin. The UV (ultra-violet) light produced by all types of arc welding can produce severe skin burns in as little as 5 minutes. A typically overlooked area of exposed skin can be the V formed between open coveralls and a lowered welding helmet: usually the operator does not realize the severity of the burn produced until it is too late.

WELDING SAFETY

**EYE PROTECTION**

➢ Select appropriate eye protection. All welding and cutting processes produce UV radiation which can damage eyesight. Goggles must be used for oxy-acetylene and propane cutting and welding processes. The goggles must be equipped with a filter shade graded as #4 or #5. Select the darkest, that is, the highest grade number, that enables you to see properly while welding or cutting. All arc welding processes require a welding helmet equipped with the correct filter shade. The filter selected should suit your eyesight. Filter grades ranging from #9 up to #13 are available. Start with a #12 or #13 grade filter and only resort to using lower filter grade numbers if you have difficulty seeing with the darker grade filter.

**BREATHING APPARATUS**

➢ Use breathing apparatus when welding in confined spaces (inside containers, tankers, vans and pressure vessels) and/or with materials that give off toxic fumes. Examples of materials that can give off toxic fumes are: aluminum alloys, copper, some stainless steels and anything galvanized. Breathing apparatus can range from simple filtration devices to oxygen packs. Note that the air filter devices used by tank cleaners and inspectors are NOT sufficient to support the oxygen requirements of a welder inside a tank.

**EAR PLUGS**

➢ Use ear plugs when in a fabricating environment. The noise produced by metal handling can damage hearing over time, especially during such processes as buck riveting.

**FLAMMABLE ITEMS**

➢ Close pockets and remove flammable items from them. A compressed liquid butane cigarette lighter is highly explosive and contains the heat energy of a standard stick of dynamite: one drop of slag can melt the plastic container and release this explosive force. NEVER use butane lighters to ignite torches.

# SHOP SAFETY

SAFE SHOP ENVIRONMENT

Shop safety requires that the welder has an awareness of the safety of those persons working around the weld area and of the environment. The materials used in welding processes have the potential to cause considerable property and environmental damage if not handled properly. Handling and properly locating welding equipment when not in use is not just the business of the service facility but also that of worker safety watchdog bodies and fire departments. The financial consequences of safety and environmental infractions can be severe and extend to closing down a business. Here are some guidelines to operating a safe shop environment:

Maintain a clean and organized work area: clutter can cause accidents.

ADEQUATE VENTILATION

> Ensure adequate ventilation: exhaust extractors and snorkel filters should be used when welding mild steel. Welding of special metals including aluminum alloys, copper, galvanized steel and many stainless steels requires snorkel extraction. Many gases produced when metals are vaporized during welding processes are toxic.

FIRE EXTINGUISHER

> Keep a fire extinguisher close by: note the location and type of the nearest fire extinguisher when welding or cutting.

FLAMMABLE MATERIALS

> Check for the proximity of explosive and flammable materials in the weld area. Remember that the petro-plastic conduit used on most copper wire in vehicle applications is not only flammable but extremely difficult to extinguish once ignited. Keep flammable fluids and materials away from the weld area: be especially careful to check the load content in truck and trailer shops.

Remember that empty flammable liquid tankers are much more dangerous than full tankers. NEVER weld near any unit with flammable placards. NEVER remove placards until the unit has been steamed.

**ELECTRICAL INTEGRITY**

➤ Try to avoid working in wet weather or damp floors. This can cause problems with shielding gases, electrode coatings and electrical integrity of arc welding stations.

# EQUIPMENT SAFETY

**EQUIPMENT STORAGE**

➤ Properly store oxy-acetylene equipment when not in use. This means having an established location identified to the fire department. In the event of a fire, it is more important to fire departments that they know the location of oxygen cylinders than acetylene. Because of the ability of oxygen to intensify any kind of combustion and the high storage pressure of oxygen, these cylinders represent an extreme hazard in the event of fire.

**EQUIPMENT MAINTENANCE**

➤ Maintain welding equipment. For oxy-acetylene equipment this means routinely inspecting hoses, gauges, cylinders, valves and torches. For arc welding, inspect the cables, electrode holders, connections and welding machines and their power sources.

**PROTECT WORK AREA**

➤ Protect others from radiated UV produced by welding. Canvas welding screens should be used to protect others from the effects of direct and indirect welding arc light. In service facilities with light colored walls, surround the welding area with screen. Aluminum and stainless steel skins on trailers are also capable of radiating indirect arc light so cover with canvass screens.

WELDING SAFETY

Courtesy of BOC GASES,
Division of BOC Canada Limited

GROUND ATTACHMENTS

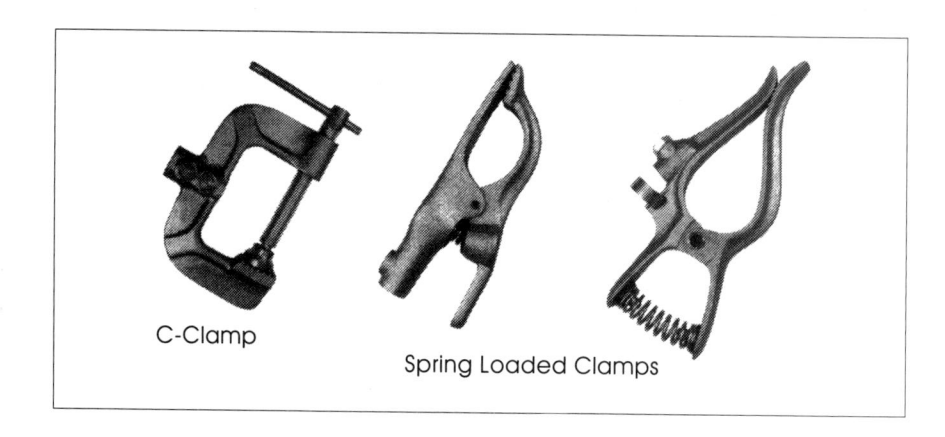

C-Clamp     Spring Loaded Clamps

ELECTRODE HOLDERS

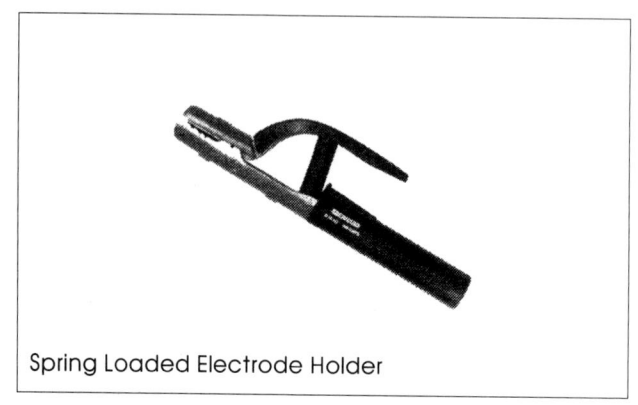

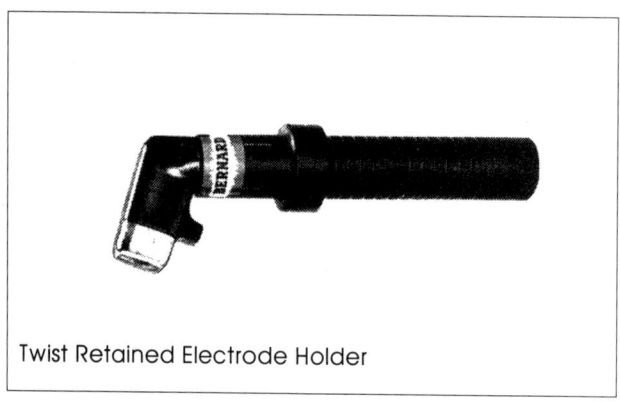

Spring Loaded Electrode Holder

Twist Retained Electrode Holder

# WELDING HELMETS

Always use the maximum filter shade number that permits you to see properly.

## Types of Welding Helmets

➢ Hand held
➢ Fixed helmet with a 2x4.25 or 4.5x5.25 lens
➢ Flip up helmet with a 2x4.25 or 4.5x5.25 lens
➢ Solar power with fixed # shade lens
➢ Solar power with a variable shade select
➢ Electronic type (requires a battery)

TYPES OF WELDING HELMETS

Courtesy of BOC GASES,
Division of BOC Canada Limited

**OPTREL ELECTRONIC WELDING HELMET**

**SOLAR CELLS**
The use of solar energy means that an on/off switch and change of battery are no longer needed.

**VISIBILITY**
Automatically darkens within a fraction of milliseconds to the correct level of protection with arc ignition.

**POTENTIOMETER KNOB**
For infinite, fine adjustment from 9 to 13, according to the welding process and the welder's degree of sensitivity.

**SENSOR BAR**
Eliminates the influences of the surrounding light. Important when welding with a low current or when welding strongly reflective surfaces. Can be detached: for example when welding in awkward positions.

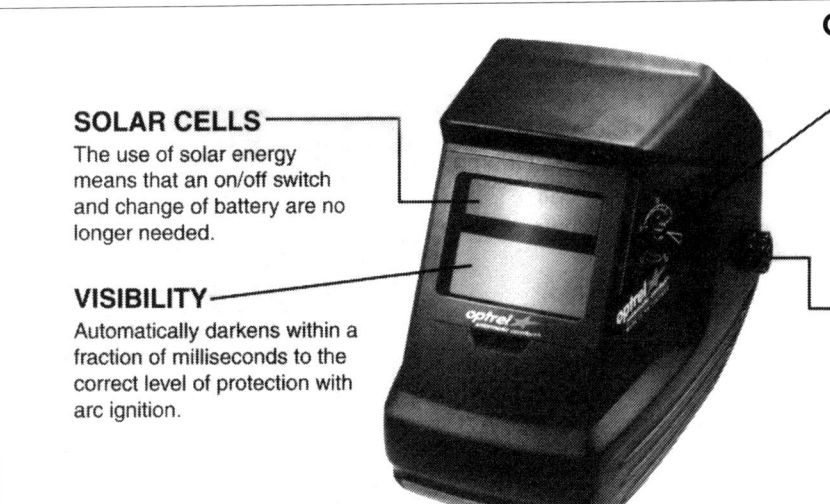

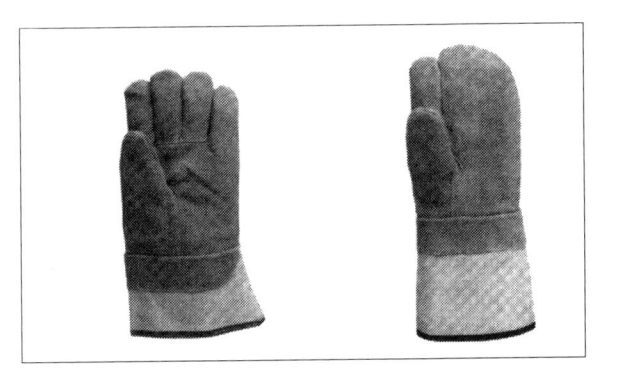

WELDING GAUNTLETS
(GLOVES)

WELDING SAFETY

# MASTER CHART OF WELDING AND ALLIED PROCESSES

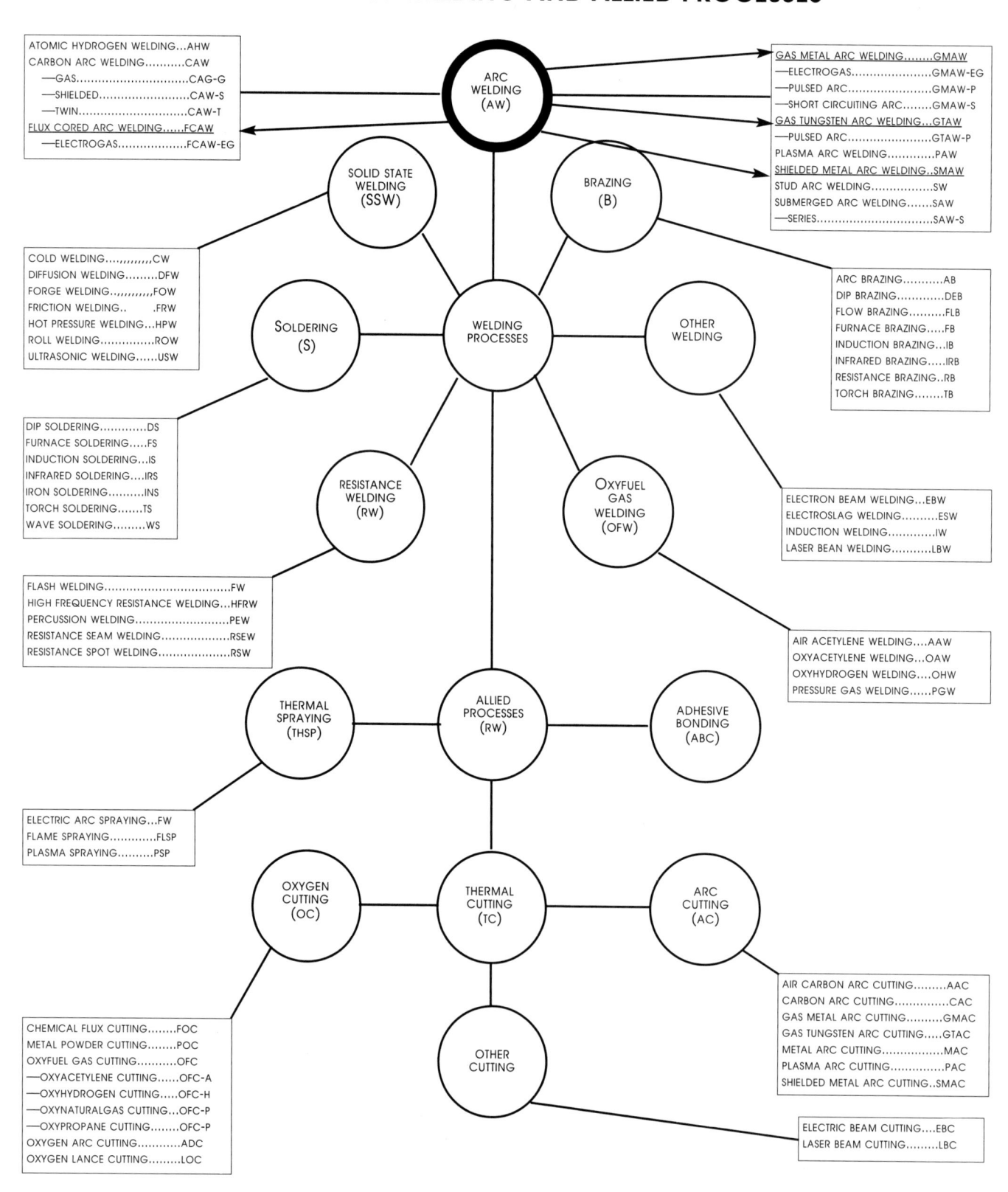

**ARC WELDING (AW)**

ATOMIC HYDROGEN WELDING...AHW
CARBON ARC WELDING...........CAW
— GAS............................CAG-G
— SHIELDED....................CAW-S
— TWIN..........................CAW-T
FLUX CORED ARC WELDING......FCAW
— ELECTROGAS..................FCAW-EG

GAS METAL ARC WELDING........GMAW
— ELECTROGAS....................GMAW-EG
— PULSED ARC.....................GMAW-P
— SHORT CIRCUITING ARC........GMAW-S
GAS TUNGSTEN ARC WELDING...GTAW
— PULSED ARC.....................GTAW-P
PLASMA ARC WELDING.............PAW
SHIELDED METAL ARC WELDING..SMAW
STUD ARC WELDING..................SW
SUBMERGED ARC WELDING........SAW
— SERIES............................SAW-S

**SOLID STATE WELDING (SSW)**

COLD WELDING..................CW
DIFFUSION WELDING..........DFW
FORGE WELDING................FOW
FRICTION WELDING............FRW
HOT PRESSURE WELDING....HPW
ROLL WELDING..................ROW
ULTRASONIC WELDING........USW

**BRAZING (B)**

ARC BRAZING............AB
DIP BRAZING.............DEB
FLOW BRAZING..........FLB
FURNACE BRAZING.....FB
INDUCTION BRAZING...IB
INFRARED BRAZING....IRB
RESISTANCE BRAZING..RB
TORCH BRAZING.........TB

**SOLDERING (S)**

DIP SOLDERING...............DS
FURNACE SOLDERING......FS
INDUCTION SOLDERING....IS
INFRARED SOLDERING.....IRS
IRON SOLDERING............INS
TORCH SOLDERING.........TS
WAVE SOLDERING...........WS

**WELDING PROCESSES**

**OTHER WELDING**

ELECTRON BEAM WELDING...EBW
ELECTROSLAG WELDING......ESW
INDUCTION WELDING..........IW
LASER BEAN WELDING........LBW

**RESISTANCE WELDING (RW)**

FLASH WELDING.....................................FW
HIGH FREQUENCY RESISTANCE WELDING...HFRW
PERCUSSION WELDING.........................PEW
RESISTANCE SEAM WELDING..................RSEW
RESISTANCE SPOT WELDING..................RSW

**OXYFUEL GAS WELDING (OFW)**

AIR ACETYLENE WELDING....AAW
OXYACETYLENE WELDING...OAW
OXYHYDROGEN WELDING....OHW
PRESSURE GAS WELDING......PGW

**THERMAL SPRAYING (THSP)**

ELECTRIC ARC SPRAYING...FW
FLAME SPRAYING.............FLSP
PLASMA SPRAYING...........PSP

**ALLIED PROCESSES (RW)**

**ADHESIVE BONDING (ABC)**

**OXYGEN CUTTING (OC)**

CHEMICAL FLUX CUTTING........FOC
METAL POWDER CUTTING........POC
OXYFUEL GAS CUTTING............OFC
— OXYACETYLENE CUTTING......OFC-A
— OXYHYDROGEN CUTTING......OFC-H
— OXYNATURALGAS CUTTING...OFC-P
— OXYPROPANE CUTTING........OFC-P
OXYGEN ARC CUTTING............ADC
OXYGEN LANCE CUTTING.........LOC

**THERMAL CUTTING (TC)**

**ARC CUTTING (AC)**

AIR CARBON ARC CUTTING..........AAC
CARBON ARC CUTTING...............CAC
GAS METAL ARC CUTTING..........GMAC
GAS TUNGSTEN ARC CUTTING.....GTAC
METAL ARC CUTTING..................MAC
PLASMA ARC CUTTING................PAC
SHIELDED METAL ARC CUTTING..SMAC

**OTHER CUTTING**

ELECTRIC BEAM CUTTING....EBC
LASER BEAM CUTTING.........LBC

# UV (ULTRA-VIOLET) RADIATION

This type of radiated light is of much shorter wavelength, just below the x-ray range, and is very penetrating. Exposure of the skin to ultra-violet radiation can cause burns similar to sunburn. Because this type of radiation is produced at high intensity, skin should never be directly exposed to it. Its effect can be to burn skin and eyes often with very serious results.

UV RADIATION

The severity of burns caused by welding UV depends upon a number of variables. These variables are:

SEVERITY OF BURNS

➢ Distance from the arc—the greater the distance, the lesser the effect.

➢ Current values—the lower the current, the weaker the emitted radiation.

➢ Sunlight or shade—if in bright sunshine the eye adjusts to the brightness and this can lessen the effect of welding UV radiation.

➢ Short exposure—the less time exposed, generally the less the damage.

Caution!!
Never focus on a welding arc with unprotected eyes even if the welding is being performed at some distance away.

PROTECT EYES

Exposure to welding arc can burn the retina of the eye. The effect on the eye is called eye flash, arc-eye, and other similar names. It is often very painful and irritating, and may last from 24 to 48 hours. Like sunburn the effects usually disappear. However, like any other burn it may leave a scar and continued exposure to severe arc flash may cause permanent eye damage. It is therefore important that welders and those working in the area use protective clothing, screens, and head "screens" or helmets with a suitable filter lens in them. The use of sunglasses or gas welding goggles will not protect the eyes, although they will reduce the effect.

EXPOSURE TO ARC FLASH

WELDING SAFETY

TREATMENT—CONSULT YOUR DOCTOR

Treatment
Eye washes, drops, or soaked pads on the eyes may reduce the pain.

Some are as follows (consult your doctor before using):

➢ Boric acid eye wash.

➢ 2% Butyl solution (anesthetic) eye drops – two treatments are usually enough.

➢ Applying a wet tea compress.

In severe cases it may be necessary to see a doctor.

ARC EYE

Caution!!
Rookie welders seldom get arc-eye. Experienced welders are far more likely to take risks when exposed to arc welding. Always regard arc welding equipment with respect. Most older welders have to use welding filters with a lower degree of UV protection because their eyesight has been damaged over the years.

# HAND TOOLS

BASIC HAND TOOLS

## Basic Tools
The basic hand tools listed below are required to remove slag and to hold hot material:

➢ Chipping hammer

➢ Wire brush

➢ Pliers

➢ Tongs

GRINDERS

## Grinders
Electric or air powered grinders are used to prepare metal pieces before welding. Most welding operations require some preparation. After welding the grinder is used to remove slag and smooth out the surface of weld for better appearance.

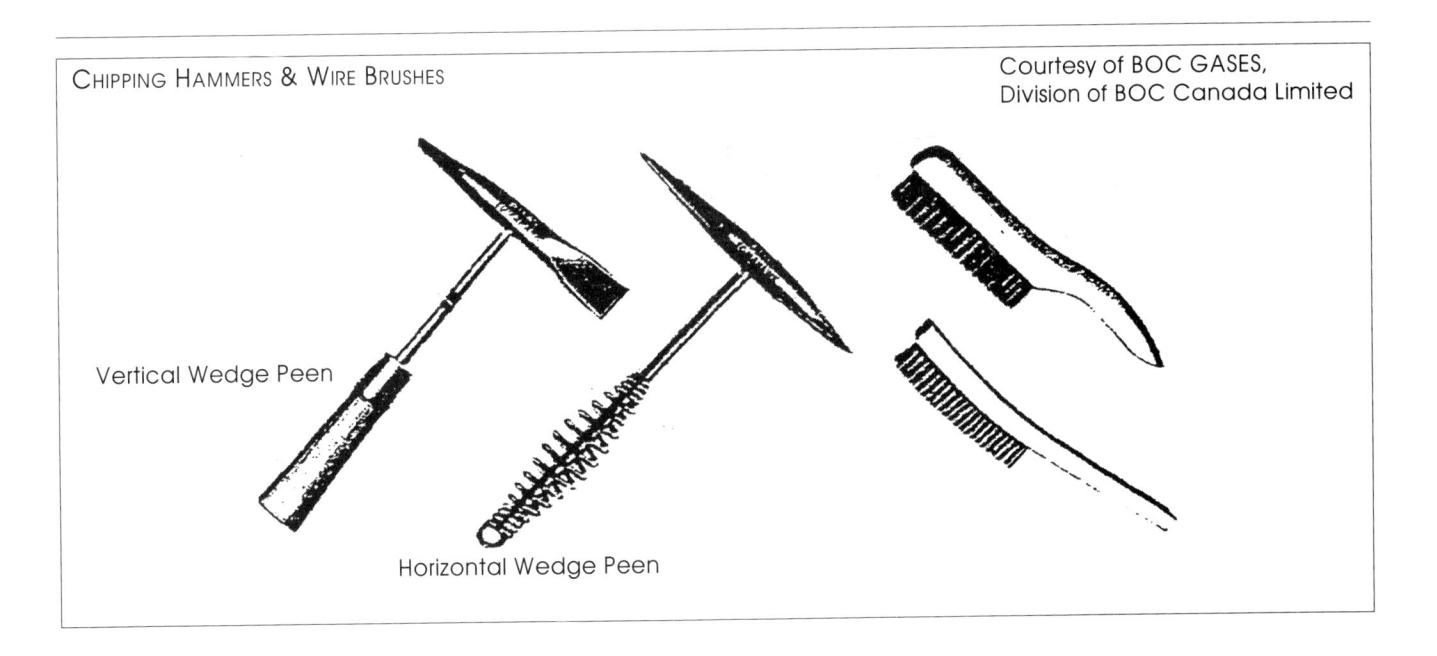

**Chipping Hammers & Wire Brushes**  
Courtesy of BOC GASES, Division of BOC Canada Limited

Vertical Wedge Peen

Horizontal Wedge Peen

## Layout Tools
To ensure a proper fit when fabricating always use layout tools such as a tape measure, set square, right angle square, combination set, protractor, center punch, scriber, ballpeen hammer and soapstone (chalk).

# FIRE CATEGORIES

All shops must be equipped with a variety of fire extinguishers. A fire extinguisher is categorized by the type of fire it is capable of extinguishing.

There are 4 categories of fire:

### ➢ Class A
Combustible materials such as wood, paper, textiles. Extinguished by: cooling, quenching, and oxygen deprivation.

### ➢ Class B
Flammable liquids, oils, grease, fuels and paints. Extinguished by: smothering (oxygen deprivation).

WELDING SAFETY

CLASS C CATEGORY

> **Class C**
> Fires that occur in the vicinity of electrical equipment, perhaps caused by a current overload.
> Extinguished by: shutting down power switches smothering with a non-conducting liquid or gas.

CLASS D CATEGORY

> **Class D**
> Combustible metals such as magnesium and sodium.
> Extinguished by: smothering with an inert chemical powder.

## TYPES OF FIRE EXTINGUISHERS

SODA-ACID

> **Soda-acid**
> Consists of bicarbonate of soda and sulfuric acid
> Uses: Type A fires only.
> Not a suitable fire extinguisher for a garage.

WATER

> **Water**
> Consists of pressurized water.
> Uses: Type A fires only.
> Not a suitable fire extinguisher for a garage.

CARBON DIOXIDE

> **Carbon Dioxide**
> Consists of compressed carbon dioxide.
> Uses: Type B and C fires — not so effective for type A fires.
> This type of fire extinguisher has uses in the shop beyond putting fires out: it can also provide a safe means of killing a runaway engine without damaging it.

DRY CHEMICAL

> **Dry Chemical**
> Consists of mostly sodium bicarbonate.
> Uses: Type A, B, C and D fires.
> Most shops should be equipped with dry chemical fire extinguishers. They are best used by directing the stream at the base of the fire and then upwards.

# SUPPRESSING FIRES

Some basic fire suppression training will equip an individual with the skills required first to assess the extent of a fire and next to handle a fire extinguisher. Fire departments and fire fighting equipment suppliers will provide training at the most basic level in how to assess the seriousness of a fire and how to extinguish small fires safely.

At least a percentage of employees in a service facility should be trained in basic fire safety. In shops with a health and safety committee, one of the functions of the committee is to identify potential fire hazards and rectify them. The objective of training in basic fire fighting techniques is not to subvert the role of the fire departments, but rather to do everything possible in a safe manner to control a fire until the arrival of fully trained firefighters.

BASIC FIRE SUPPRESSION TRAINING

# OXY-ACETYLENE PROCESSES    2

Oxy-acetylene is widely used in the transportation industry and even if a technician never uses a set of cutting or welding torches, some basic safety instruction in welding gas safety is required.

WELDING GAS SAFETY

Technicians use oxy-acetylene for heating and cutting on a daily basis. Less commonly this equipment is used for braising and welding. Some basic instruction in the nature of oxy-acetylene equipment safety and handling is required. The following information should be known by every person using and working in close proximity with oxy-acetylene equipment.

## ACETYLENE

Acetylene is an unstable gas produced by immersing calcium carbide in water. It is stored in a compressed state, dissolved in acetone at pressures of approximately 250 psi / 1720 kPa. Acetylene cylinders are fabricated in sections and seam welded: next a paste of cement, lime silica, cement and asbestos is baked within the cylinder forming a honeycomb structure. The cylinder is then charged with liquid acetone which is capable of absorbing 25 times its own volume of acetylene at normal temperatures. The base of the acetylene cylinder is concave and has two or more fusible plugs threaded into 2 apertures. The fusible plugs are made of a lead base alloy and are designed to melt at around 100°C / 212°F. The idea is that if the cylinder were exposed to heat, the fusible plugs are designed to melt and permit the acetylene to escape thus avoiding exploding the cylinder.

ACETYLENE STORAGE

Oxy-acetylene Processes

# Oxy-acetylene Welding Equipment

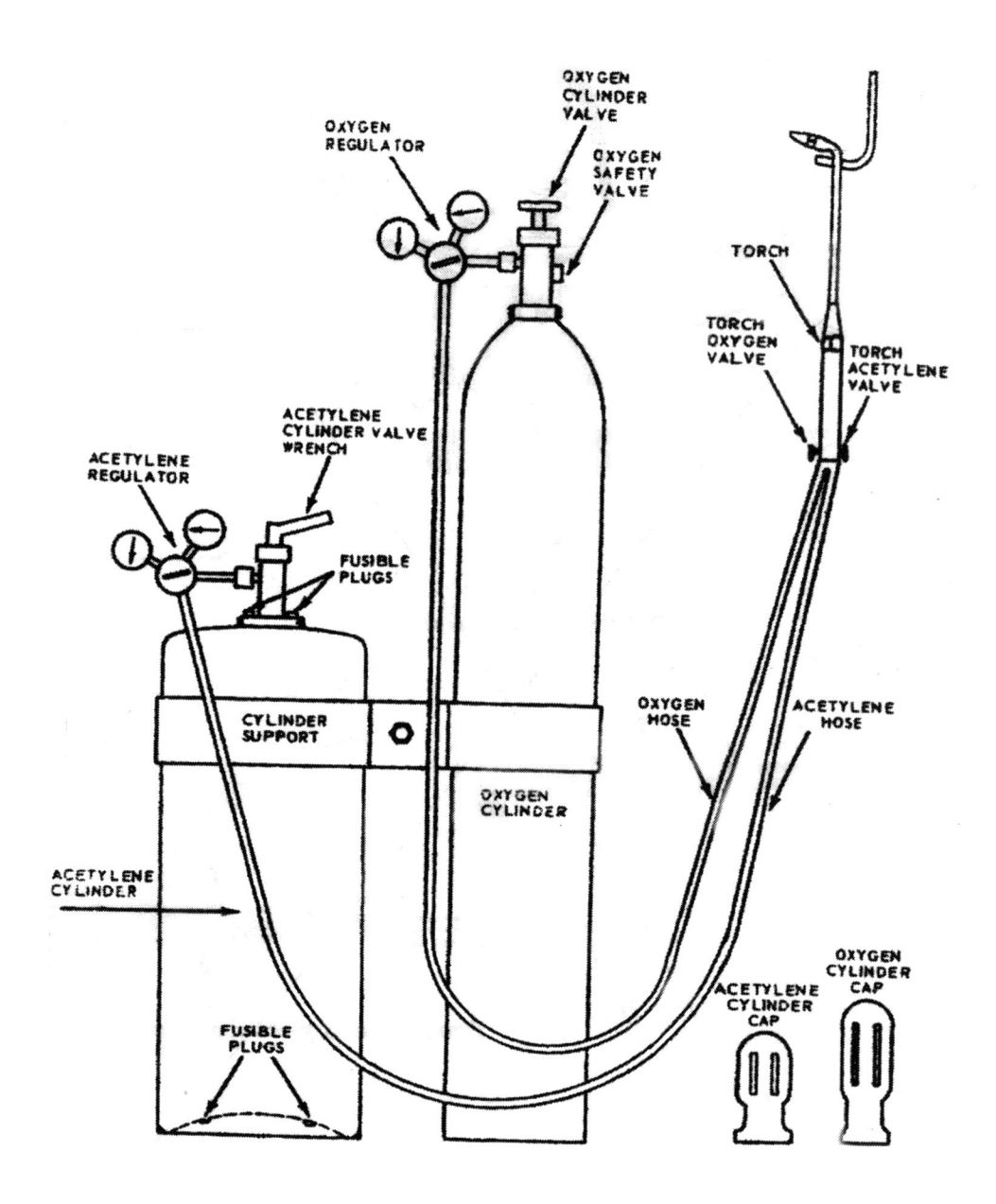

Note the location of the fusible plugs on the acetylene cylinder and that of the rupture disc/safety valve on the oxygen cylinder.

## ACETYLENE REGULATORS AND HOSE COUPLINGS

Acetylene regulators and hose couplings use a left hand thread: the regulator gauge working pressure should never be set at a value exceeding 15 psi / 100 kPa: acetylene becomes extremely unstable at pressures higher than 15 psi. The acetylene cylinder should always be used in the upright position. Using an acetylene cylinder in a horizontal position will result in the liquid acetone draining into the hoses.

## QUANTITY IN A CYLINDER

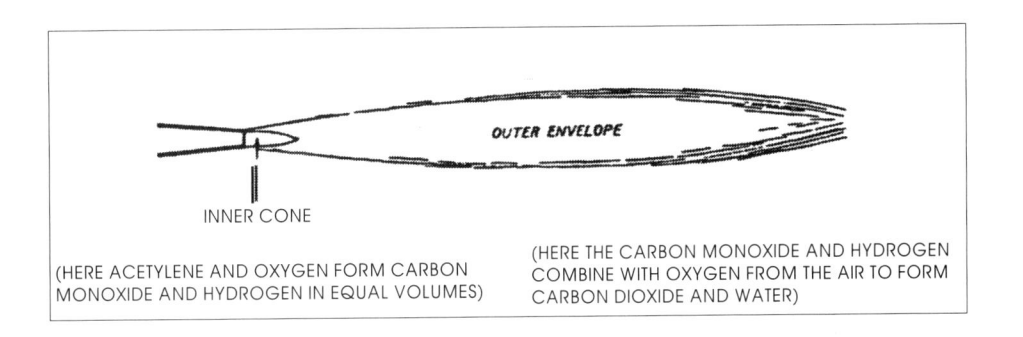

It is not possible to determine with any accuracy the quantity of acetylene in a cylinder by observing the pressure gauge because the acetylene is in a dissolved condition. The only accurate way of determining the quantity of gas in the cylinder is to weigh it and subtract this from the weight of the full cylinder, often stamped on the side of the cylinder.

OUTER ENVELOPE

INNER CONE

(HERE ACETYLENE AND OXYGEN FORM CARBON MONOXIDE AND HYDROGEN IN EQUAL VOLUMES)

(HERE THE CARBON MONOXIDE AND HYDROGEN COMBINE WITH OXYGEN FROM THE AIR TO FORM CARBON DIOXIDE AND WATER)

# OXYGEN

## OXYGEN CYLINDERS

Oxygen cylinders are forged in a single piece, no part of which is less than 1/4" / 6.4mm thick: the steel used is armor plate quality, high carbon steel suitable for pressure vessels. Oxygen is contained in the cylinder at a pressure of one ton per square inch so the design is consistent with what is required for a high pressure vessel with radial corners. Oxygen cylinders are required to be periodically hydrostatic tested at 3,300 psi. The safety device on an oxygen cylinder is a rupture disc designed to burst if cylinder pressure exceeds its normal value such as when exposed to fire. It should be noted that oxygen cylinders tend to pose more problems than

## Oxy-acetylene Processes

acetylene when exposed to fire. They should always be stored in the same location in a service shop when not in use (this should be identified to the Fire Department during an inspection) and never left randomly on the shop floor.

**Oxygen Regulator and Hose Fittings**

Oxygen regulator and hose fittings use a right hand thread. The cylinder pressure gauge will accurately indicate the oxygen quantity in the cylinder, meaning that the volume of oxygen in the cylinder is approximately proportional to the pressure.

**Oxygen Storage Cylinders**

Oxygen is stored in the cylinders at a pressure of 2,200 psi / 15 MPa and the hand wheel actuated valve, forward-seats to close the flow from the cylinder and back-seats when the cylinder is opened. It is important to ensure that the valve is fully opened when in use. The consequence of not fully opening the valve is leakage past the valve threads.

Courtesy of BOC GASES, Division of BOC Canada Limited

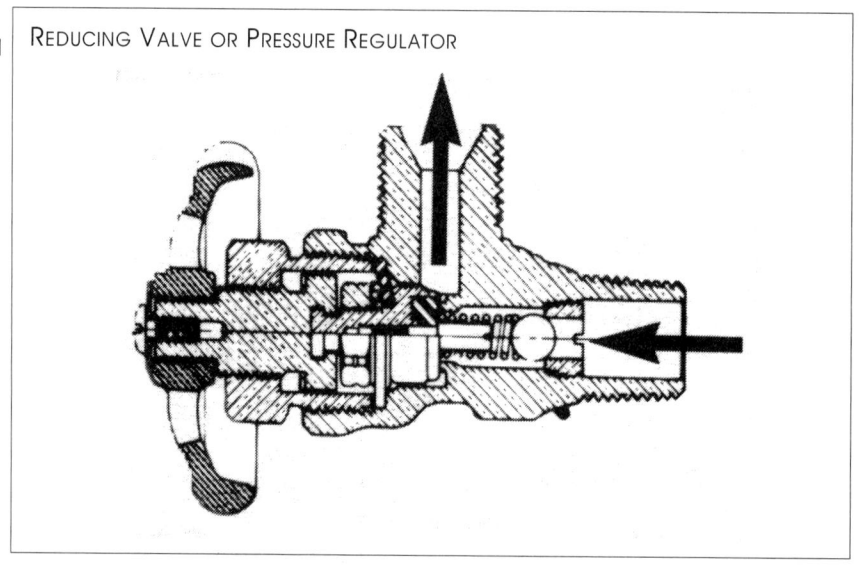

REDUCING VALVE OR PRESSURE REGULATOR

# REGULATORS AND GAUGES

A regulator is a device used to reduce the pressure at which gas is delivered: it sets the working pressure of the oxygen or fuel. Both oxygen and fuel regulators function similarly in that they increase the working pressure when turned clock-wise They close off the pressure when backed out counter clock-wise.

Pressure regulator assemblies are usually equipped with two gauges. The cylinder pressure gauge indicates the actual pressure in the cylinder. The working pressure gauge indicates torch working pressure. This should be trimmed using the regulator valve to the required value while under flow.

# HOSES AND FITTINGS

The hoses used with oxy-acetylene equipment are usually color coded. Green is used to identify the oxygen hose and red identifies the fuel hose. Each hose connects the cylinder regulator assembly with the torch. Hoses may be single or paired (Siamese). Hoses should be routinely inspected and replaced when defective. A leaking hose should never be repaired by wrapping with tape. In fact, it is generally bad practice to consider repairing welding gas hoses by any method: simply replace them when they fail.

Fittings couple the hoses to the regulators and the torch. Each fitting consists of a nut and gland. Oxygen fittings use a right hand thread and fuel fittings use a left hand thread. The fittings are machined out of brass which has a self lubricating characteristic. Never lubricate the threads on oxy-acetylene fittings.

# BACKFIRE

Backfire is a condition where the fuel ignites within the nozzle of the torch producing a popping or squealing

OXY-ACETYLENE PROCESSES

noise: it often occurs when the torch nozzle overheats. Extinguish the torch and clean the nozzle with tip cleaners. Torches may be cooled by immersing in water briefly with the oxygen valve open.

## FLASHBACK

FLASHBACK

Flashback is a much more severe condition than backfire: it takes place when the flame travels backwards into the torch to the gas mixing chamber and beyond. Causes of flashback are inappropriate pressure settings (especially low pressure settings) and leaking hoses/fittings. When a backfire or flashback condition is suspected, close the cylinder valves immediately, beginning with the fuel valve. Flashback arresters are usually fitted to the torch and will limit the extent of damage when a flashback occurs.

## TORCHES AND TIPS

IGNITING AND EXTINGUISHING TORCHES

Torches should be ignited by first setting working pressure setting under flow for both gases, then opening the fuel valve only and igniting the torch using a flint spark lighter. Set the acetylene flame to a clean burn (no soot), then open the oxygen valve to set the appropriate flame. When setting a cutting torch, set the cutting oxygen last. When extinguishing the torch, close the fuel valve first, then the oxygen: next the cylinders should be shut down and finally the hoses purged.

WORKING PRESSURES FOR TIPS

Welding, cutting and heating tips may be used with oxy-acetylene equipment. Consult a welder's manual to determine the appropriate working pressures for the tip/process to be used. There is a tendency to set gas working pressure high. Even when using a large heating tip often described as a rosebud, the working pressure of both the acetylene and the oxygen should be set at no more than 7 psi. /50 kPa.

# EYE PROTECTION

Safety requires that a #4 - #6 grade filter be used whenever using an oxy-acetylene torch. The flame radiates ultra-violet light that can damage eyesight.

# OXY-ACETYLENE PRECAUTIONS

- Store oxygen and acetylene upright in a well ventilated, fire-proof room.
- Protect cylinders from snow, ice and direct sunlight.
- Remember that oil and grease may ignite spontaneously in the presence of oxygen.
- Never use oxygen in place of compressed air.
- Avoid bumping and dropping cylinders.
- Keep cylinders away from electrical equipment where there is a danger of arcing.
- Never lubricate the regulator, gauge, cylinder or hose fittings with oil or grease.
- Blow out cylinder fittings before connecting regulators: make sure the gas jet is directed away from equipment and other people.
- Use soapy water to check for leaks: NEVER use a flame to check for leaks.
- Thaw frozen spindle valves with warm water, NEVER a flame.

# ADJUSTMENT OF THE OXY-ACETYLENE FLAME

To adjust an oxy-acetylene flame, the torch acetylene valve is first turned on and the gas ignited. At the point of ignition, the flame will be yellow and producing black smoke. The acetylene pressure should next be increased using the torch fuel valve.

OXY-ACETYLENE PROCESSES

This will increase the brightness and reduce the smoking. At the point the smoking disappears, the acetylene working pressure can be assumed to be correct for the nozzle jet size used.

Next, the torch oxygen valve is turned on. This will cause the flame to become generally less luminous and an inner blue luminous cone surrounded by a white colored plume will form at the tip of the nozzle. The white colored plume indicates excess acetylene. As more oxygen is supplied, this plume reduces until there is a clearly defined blue cone with no white plume visible. This indicates the neutral flame used for most welding and cutting operations.

OXIDIZING FLAME

If, after setting a neutral flame, the oxygen supply is increased, the blue cone will become smaller and sharper in definition and the outer envelope will become streaky. This is known as an oxidizing flame indicating excess oxygen. For most welding procedures an oxidizing flame should be avoided, but in some special applications, a slightly oxidizing flame may be recommended.

CARBURIZING FLAME

A carburizing flame is indicated by the presence of a white plume surrounded inner blue cone. A carburizing flame should be avoided in most welding operations, although a very slightly carburizing flame is used in certain special applications.

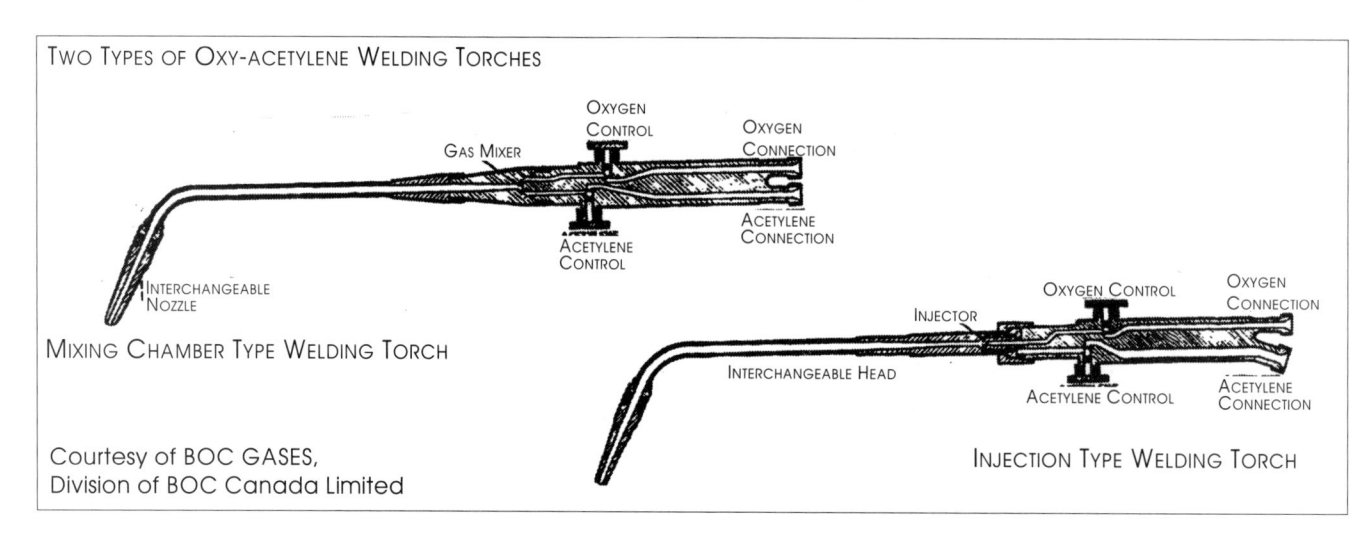

TWO TYPES OF OXY-ACETYLENE WELDING TORCHES

MIXING CHAMBER TYPE WELDING TORCH

INJECTION TYPE WELDING TORCH

Courtesy of BOC GASES,
Division of BOC Canada Limited

# OXY-ACETYLENE WELDING PRACTICE

*GAS WELDING OPERATION TECHNIQUES*

No amount of text can substitute for the actual practice of welding. After some initial demonstration, the learner welder should run as many welds as possible. Welding skill is generally the result of practice and acquiring some understanding of what constitutes a good weld.

Generally, a right handed, right dominant eye welder, will operate gas welding equipment working from the right towards the left. The torch is held in the right hand and the filler wire in the left hand. The torch angle should be typically 60 to 70 degrees to the work. The closer to vertical the torch angle, the higher the heat delivered to the base metal: this means that the welder can use torch angle to control the amount of heat (and penetration). The torch may be moved with a gentle side-to-side or circular movement. The distance of the torch to the work can also be used (along with torch angle) to control the amount of heat in the base metal.

The filler wire is held at a more acute angle, usually 30 to 40 degrees and dipped into the molten puddle. The process of adding filler wire to the molten puddle produced by the torch, should be one of rhythmic dipping into the puddle and pulling away to an inch or two distant. The wire should NEVER be melted by the torch flame rather than by dipping into the molten puddle or the result will be poor penetration. When performing positional welds, the filler wire should be dipped into the highest point of the puddle to counter the effect of gravity on the molten puddle.

Left handed, left eyed welders should generally reverse the above outlined procedure. Welders who have cross laterality (the dominant eye is different from the dominant hand) should begin by using the method appropriate for the dominant eye—and be prepared to experiment with both methods until they determine which is going to be more comfortable.

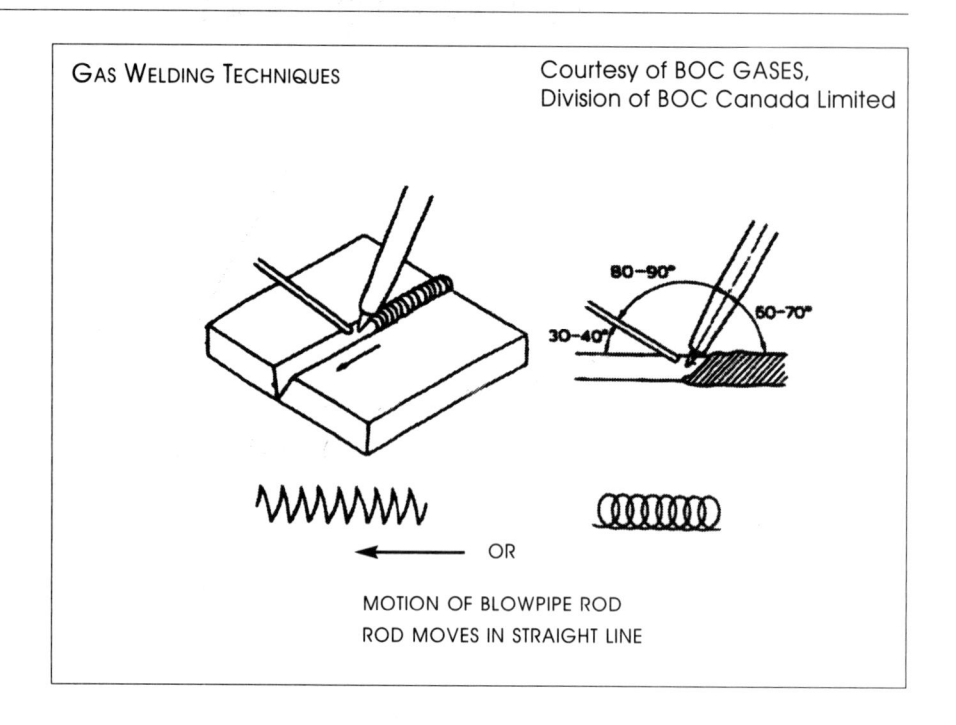

## Oxy-Acetylene Processes

**Gas Welding Techniques** — Courtesy of BOC GASES, Division of BOC Canada Limited

MOTION OF BLOWPIPE ROD
ROD MOVES IN STRAIGHT LINE

# GAS CUTTING OF STEEL     3

*CUTTING METHODS*

Iron and steel may be cut by oxy-hydrogen, oxy-propane, oxy-natural gas and oxy-acetylene cutting torches. There are also a number of arc cutting methods which will not be covered in this section. In most heavy equipment, auto and truck service shops, the oxy-acetylene cutting process is used and, less often, oxy-propane.

*TWO-STAGE OPERATION*

There are two stages to any gas cutting operation. First, a heating flame is directed on the steel to be cut and raises its temperature to a bright red heat. Next, at the point where the required temperature in the base metal is achieved, a stream of high pressure oxygen is directed onto the hot metal. This results in the iron being immediately oxidized to iron oxide. Since the melting point of iron oxide is much lower than that of steel, the iron oxide is blown away by the pressure of the oxygen gas exiting the torch and the steel either side of the oxide is left unaffected. The heat to maintain the cut once it has been started is provided partly by the heating jet and partly by the heat of the chemical oxidation reaction.

## CUTTING TORCH

*OXY-ACETYLENE CUTTING TORCH*

The oxy-acetylene cutting torch has a mixing chamber in the head, meaning that the mixing of the acetylene and the oxygen takes place in the torch. The cutting nozzle is made with annular orifices from which the mixed oxy-acetylene gas exits. The cutting oxygen is controlled by a spring loaded lever. When this is

# Gas Cutting of Steel

## Oxy-acetylene Cutting Torch

Courtesy of BOC GASES,
Division of BOC Canada Limited

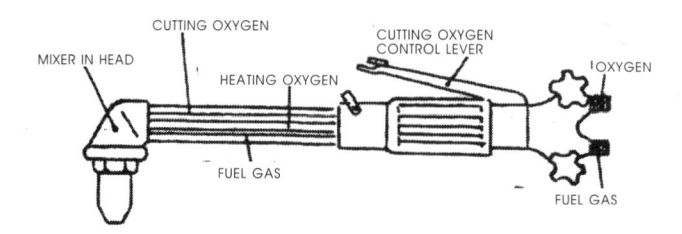

## Sectional Views of Cutting Heads

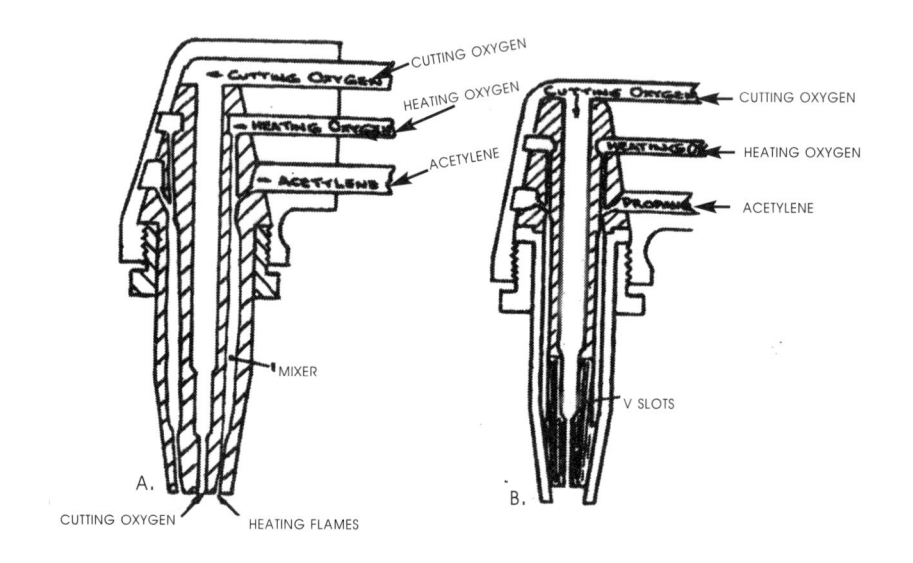

## Cutting Head for Thin Steel Sheet

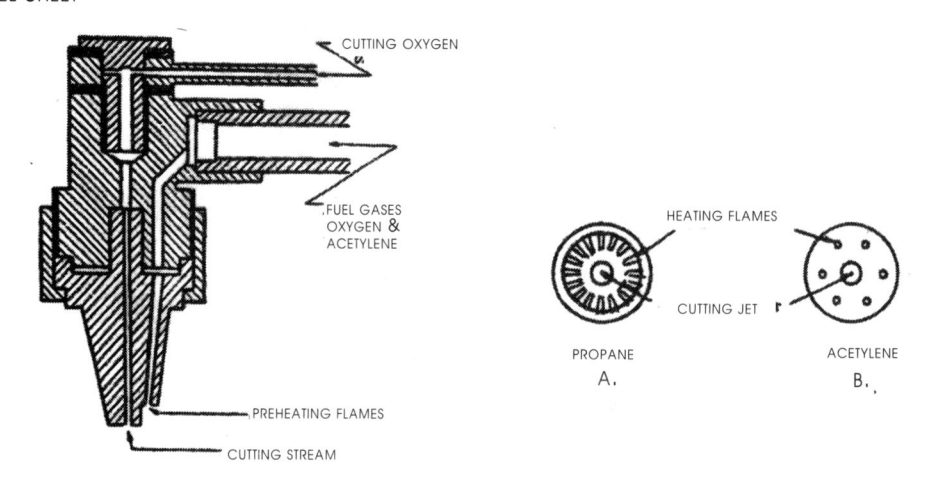

depressed by the operator, a stream of oxygen exits from a central orifice in the cutting tip. This central orifice is surrounded by the heating flame orifices.

The size of the cutting tip or nozzle used will be determined by the thickness of the metal to be cut. Using too large a cutting nozzle will have the effect of applying excessive heat to the metal being cut and result in a sloppy cut with plenty of oxidized slag around it.

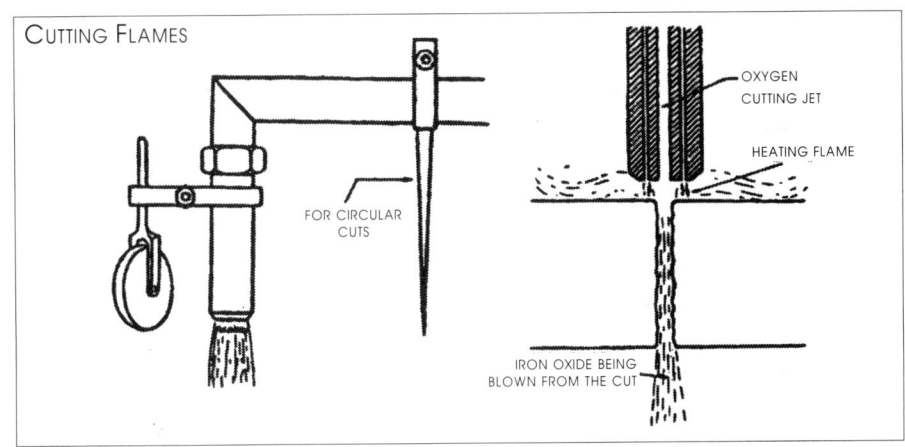

Courtesy of BOC GASES,
Division of BOC Canada Limited

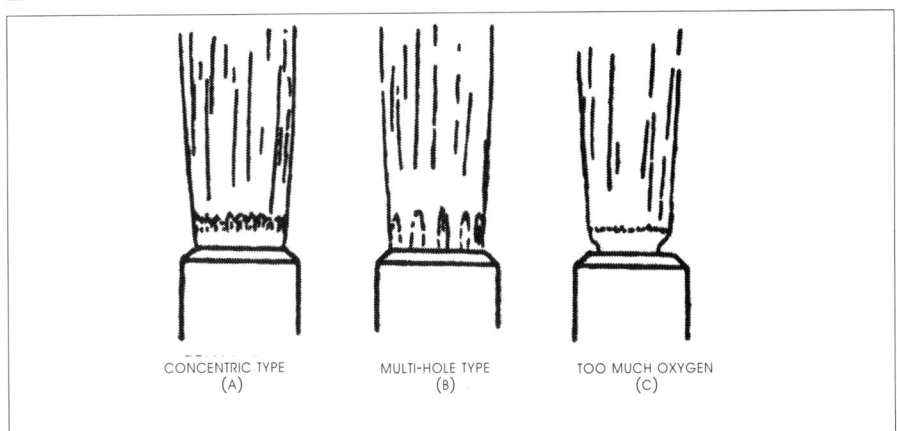

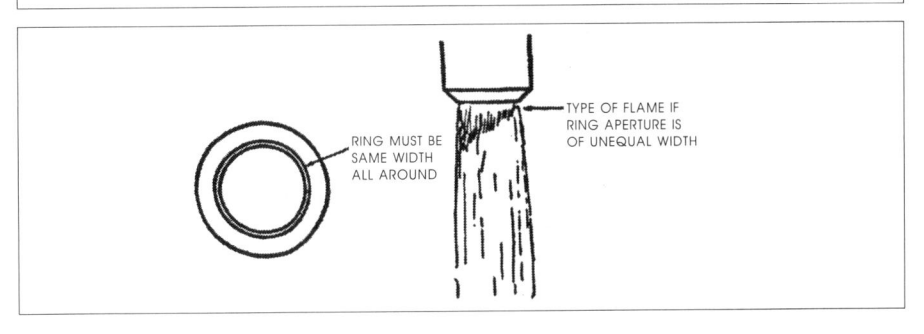

# OXY-ACETYLENE FLAME ADJUSTMENT

**FLAME ADJUSTMENT**

The flame should be adjusted by first turning on the acetylene, then igniting it and turning the fuel pressure valve to the point that it ceases to smoke. Next, open and trim the oxygen valve to produce a neutral flame. A neutral cutting flame is produced when the excess acetylene feather surrounding the blue cone exiting each orifice is removed. Too much oxygen will produce a sharp blue cone and a more pronounced torch hiss. Ensure the flame is set neutral. This is important! A slightly oxidizing or carburizing flame can cause messy-looking cuts and damage the base metal.

Next, push down the cutting lever. If a white feather is produced, it indicates excess acetylene. This is caused by decreased oxygen supplied to the heating flame when the cutting oxygen is released. Adjust the heating flame so that it is exactly neutral when the cutting lever is depressed. Now the metal to be cut can be heated and at the point it achieves bright cherry red, the cutting lever can be depressed to begin the cut.

**CUTTING NOZZLE**

The cutting nozzle must be kept clean. Nozzles clog easily with metal debris during cutting. The metal to be cut should be clean of oil and grease. The distance of the cutting torch to the metal to be cut will vary somewhat but it should be around 1/8" to 1/4" (3.2 to 6 mm). Use cutting guides and jigs whenever possible to ensure that the neatest cut is produced.

**PREPARING TO CUT**

Caution!!
Take some time preparing to cut. Make sure the flame is properly set and use cutting guides. Taking a little time preparing before a cut can save plenty of time working with a grinder rescuing a poorly cut piece of steel.

**NOT ALUMINUM**

Do not attempt to cut aluminum or aluminum alloys with oxy-acetylene. Oxides of aluminum have a higher melting point than pure aluminum. The result of attempting to cut aluminum with gas cutting torches is a nasty mess.

# ARC WELDING                                          4

## Arc WELDING PRINCIPLES

Arc welding is a fusion welding process that uses electrical current to create an electric arc, producing enough heat to melt, and physically fuse the base metal using a filler metals. The filler metal is in the form of a coated alloy electrode often called a rod. The immediate weld area is shielded from the atmospheric air by the vaporizing of the flux coating on the welding electrode. This forms a gaseous shroud around the arc and the welding puddle. The gas shroud is essential. It protects the weld area from exposure to air and the moisture contained in air. The temperatures created by an electric arc are between 6500°F and 7000°F / 3500°C and 3800°C.

Arc Welding Principles

### Polarity

A welding arc created by the flow of current consists of a flow of electrons and of molten filler wire across the air gap. Electrons flow from negative to positive. A negative charge differential is a really an excess of electrons in any point of a circuit. A positive charge is a shortage of electrons. Whenever a charge differential exists, so does the opportunity for current flow. This occurs when a circuit is closed or the charge differential is sufficient for an arc to form. Electrons flow to balance the charges. The charge differential is initiated by an arc welding station which then creates high current flow. The actual direction of the current flow depends on the polarity set at the welding station.

Selecting Polarity

# Arc Welding

Typically, current flow is routed from the welding machine through the electrode cable to the electrode holder—or oppositely, depending on the polarity selected. From the electrode holder, current flows through the electrode and across the gap between the tip of the electrode and the base metal. From the base metal, the current returns through the ground cable to the welding machine.

## Direct Current Electrode Negative (DC Straight Polarity)

**DC Straight Polarity**

When the base metal has a positive charge differential (connected to the positive terminal of the welding machine) and the welding electrode has a negative charge (connected to the negative terminal of the welding machine), the polarity is called direct current electrode negative (DCEN). The direction of electron flow is therefore from the electrode to the base metal. This means that the molten metal flows from the electrode to the base metal in same direction as electron flow. This is called straight polarity. DC straight polarity welding results in approximately two-thirds of the heat produced by the arc to occur in the work being welded and one-third in the electrode. DC straight polarity welding electrodes are used especially where good structural integrity is required.

Straight Polarity Characteristics

**Straight Polarity Characteristics**

- Electrode melts more slowly and is easier to use.
- Weld is deposited at a slower rate.
- Base metal is heated more quickly providing deep penetration.
- Used when welding large gauge metals.

## Direct Current Electrode Positive (DC Reverse Polarity)

**DC Reverse Polarity**

When the electrode has a positive charge (connected to the positive terminal of the welding machine), and the base metal has a negative charge (connected to the negative terminal of the welding machine), the electrons flow from base metal to the rod.

## DC Reverse Polarity

The molten metal still flows from the rod to the base metal, that is in reverse to electron flow. Because the molten metal travels in reverse direction to the electrons, this is known as reverse polarity. DC reverse polarity welding results in approximately two-thirds of the heat produced by the arc to occur at the electrode and one-third in the work being welded. DC reverse polarity welding electrodes are used especially where maximum strength is required.

Reverse Polarity Characteristics
- Electrode melts more rapidly.
- Used for root runs in open butt welds.
- Base metal is heated more slowly.
- Used for welding thin metals.

## Selecting Polarity

DC welding machines can function in both DCEN (straight polarity) and DCEP (reverse polarity). By simply reversing the leads at the welder, or by flipping a switch, either polarity may be selected. As we have already said, welding circuit polarity affects the weld process and the quality of the final weld, as follows:

DCEN (Straight Polarity)
- Puts more heat into the base metal.
- Produces deeper penetration.
- Often used with electrodes designed for root runs.

DCEP (Reverse Polarity )
- Causes rod to melt faster and the deposit more metal.
- Suitable for welding very thin metals.
- Often used in production welding because it is usually faster.
- Used in applications likely to crystallize if too much heat is applied to the base metal.

ARC WELDING

ALTERNATING CURRENT WELDING

### Alternating Current Welding

Mains electricity in North America is supplied at a frequency of 60 cycles per second or 60 hertz. One cycle is completed when current flows for 1/120 of a second in one direction, and then for 1/120 of a second in the opposite direction. Electrodes designed for alternating current welding produce a neutral or reducing gas shield while welding. The weld penetration of alternating current welding ranges between that of DCEP and DCEN welding, as does the heat, which is divided evenly between the base metal and the electrode.

## ARC WELDING EQUIPMENT

CONSTANT CURRENT MACHINE

### Machines

Machines used for arc welding are constant current type. This means that when voltage changes, current remains the same. The actual voltage in an active welding circuit is dependent on the length of the arc. When the arc length increases, the voltage must also increase in order to maintain the same amount of current flow. Constant current machines respond to changes in voltage accordingly. The voltage versus amperage curve is called a droop. Most electric arc stations produce an open circuit voltage (OCV) of around 70 volts and a closed circuit voltage (CCV) of around 20 volts.

RECTIFIERS

DC welding stations are known as rectifiers because they rectify the main AC supply voltage to DC. AC welding stations are transformers; specifically they are step-down transformers. DC welding machines tend to be more versatile than AC machines because of the control provided by the selection of polarity. This enables them to be able to handle positional welding and a greater range of metal thicknesses. Many DC machines also have AC capability. AC welding machines tend to cost less, but are less versatile. A mains supply AC voltage may be 110 V,

# Arc Welding

220 V, or 550 V to both AC and DC arc welding stations. Because of the high feed voltage, extreme care must be exercised when working around welding stations outside of the welding electrical circuit.

## Duty Cycles
When welding stations are used in the higher current ranges, they often subject to duty cycle limitations. Exceeding the duty cycle can cause the welding machine to overheat and may trip an internal circuit breaker. The duty cycle is based on a 10-minute time span and expressed as a percentage. Welding machines come with duty cycle charts that should be consulted to ensure welding safety and optimum weld quality. For instance, a set rated at a 60% duty cycle at 150 amps = 6 minutes welding time and 4 minutes down time when run at that amperage setting.

*Duty Cycle Charts*

## Welding Cables
A ground cable and an electrode cable are required to connect to the work piece. Cables are usually made of fine strands of copper wire. Cables are flexible and well-insulated by a protective layer of rubber. The insulation is rugged and can withstand most of the rigors of the work environment. Cables come in various sizes and a numbering system is used to standardize these sizes. The larger the sectional area of cable the greater its ability to carry current.

*Welding Cables*

For instance, welding cable that overheats quickly in use is probably not thick enough to handle the work being performed. The longer a cable is, the greater the resistance, so extra long welding cables come with a disadvantage. They must compensate for the extra length of the welding cable by having a larger diameter.

## Welding Cable Connectors
Cables may use clamped lugs on their ends that connect to studs or bolts, or quick connect couplers that push and turn onto terminals. Quick connections may not be interchangeable between manufacturers.

*Welding Cable Connectors*

## Arc Welding

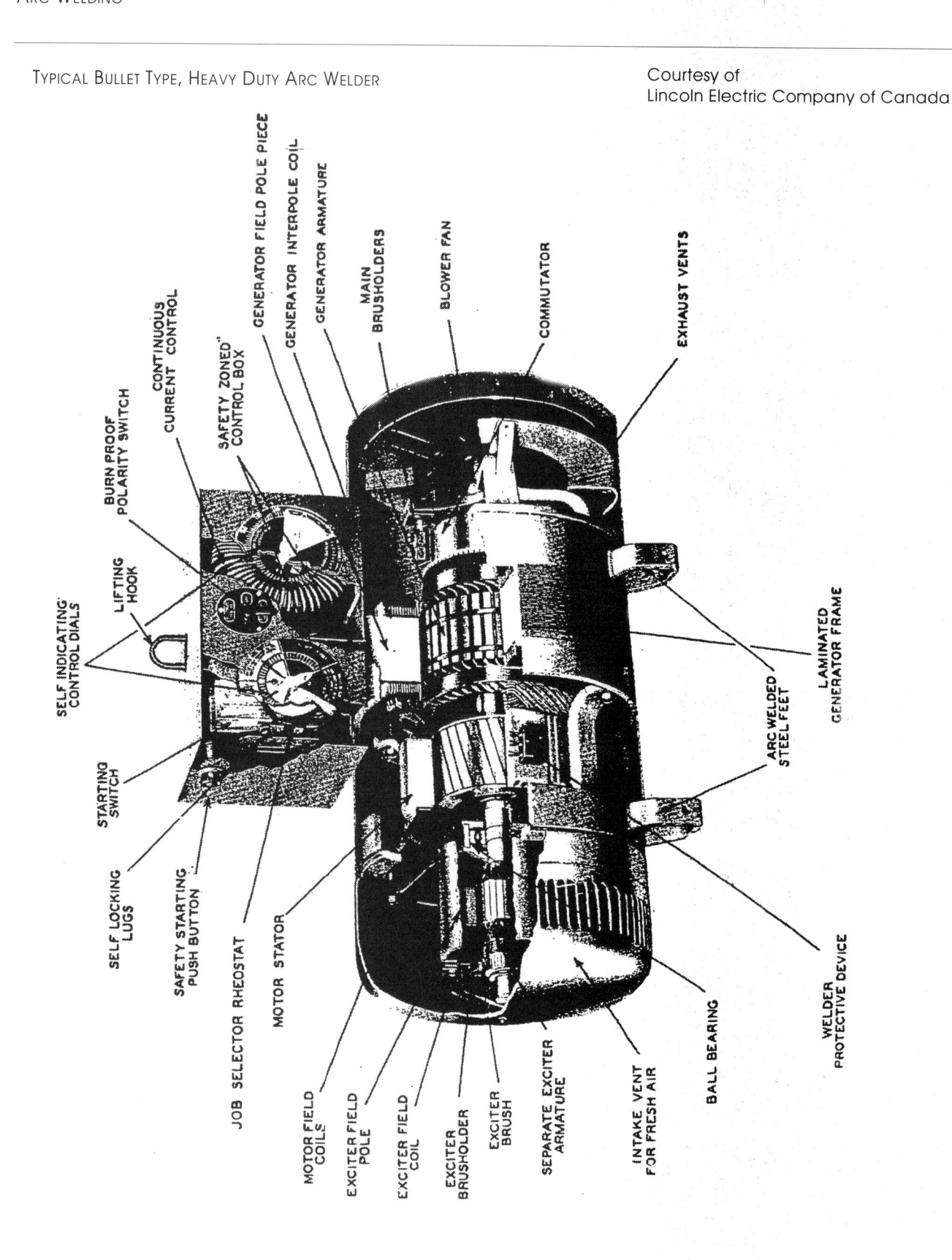

**Typical Bullet Type, Heavy Duty Arc Welder**

Courtesy of
Lincoln Electric Company of Canada

Labels (clockwise from upper left):
- Burn Proof Polarity Switch
- Continuous Current Control
- Safety Zoned Control Box
- Generator Field Pole Piece
- Generator Interpole Coil
- Generator Armature
- Main Brusholders
- Blower Fan
- Commutator
- Exhaust Vents
- Laminated Generator Frame
- Arc Welded Steel Feet
- Welder Protective Device
- Ball Bearing
- Intake Vent for Fresh Air
- Separate Exciter Armature
- Exciter Brush
- Exciter Brusholder
- Exciter Field Coil
- Exciter Field Pole
- Motor Field Coils
- Motor Stator
- Job Selector Rheostat
- Safety Starting Push Button
- Self Locking Lugs
- Starting Switch
- Lifting Hook
- Self Indicating Control Dials

## Ground Connectors

The ground end of the cable end is attached to the base material to be welded. A good electrical connection is necessary to prevent high resistance that results in poor welding control. It may also be difficult to strike the arc when the ground connection is poor. Ground connectors are commonly spring loaded clamps, but there are many varieties of specialty ground clamps and bases available.

## Electrode Holder

The electrode holder clamps the electrode and is designed to comfortably fit in the welder's hand. It has a well-insulated handle and a clamp with copper alloy jaws to grip the bare of flux end of the welding rod and provide good electrical contact. The electrode holder is designed so that there is no external part that can short to the bare metal.

# ELECTRODES

Welding electrodes are solid cylindrical metal rods, coated with flux. The wire or rod is an alloy of the metal to be welded, or in some instances, pure metal. During welding, the wire melts and serves as filler metal. The flux coating is designed to melt and/or vaporize to a gas that shields the arc and weld puddle from ambient air. Welding electrodes sold in North America are made to conform to standards developed by the American Welding Society. The standards are revised and published every 5 years.

## Electrode Coatings

Flux chemicals are combined in exact proportions, mixed into a clay-like consistency and applied to the rod in precise thickness.

The coatings serve several functions that may include some or all of the following:

➢ Adding filler metal: iron is commonly added to improve the metallurgical characteristics of the weld.

ARC WELDING

| | |
|---|---|
| COATING FUNCTIONS | ➢ Float weld impurities out of the weld puddle to form the slag coating. |
| | ➢ Provide a hard slag coating to slow cooling of weld. |
| | ➢ Change chemical properties of metal cooling through alloying. |
| SHIELDING GASES | Shielding gases created by the burning of the rod covering may be neutral or reducing gases. |
| NEUTRAL GASES | **Neutral Gases**<br>Neutral gases do not change the chemical properties of the weld. |
| REDUCING GASES | **Reducing Gases**<br>Reducing gases reduce the level of oxygen in the surrounding atmosphere through chemical reaction. |
| | Low hydrogen electrodes provide a shielding gas that prevents hydrogen from combining chemically with molten metal, which would weaken it. |
| ELECTRODE SPECIFICATIONS | The composition of the electrode covering may work with DCEN only, DCEP only, or either setting. Some welding rods are designed to function with AC exclusively and a few will work with all polarities and current options. Check the manufacturer specifications for the electrodes you are using to determine suitability for the equipment and task at hand. |
| ELECTRODE SIZES | **Electrode Sizes**<br>The size of an electrode refers to the size of the wire, without covering. |
| | Standard lengths are 9"/23cm, 12"/30cm, 14"/35cm and 18"/46cm. |
| | Standard wire sizes or diameters include 1/16" (1.5mm), 5/64" (2.0mm), 3/32" (2.4mm), 1/8" (3.2mm), 5/32" (4.0mm), 3/16" (4.8mm), 7/32" (5.5mm), 1/4" (6.4mm), 5/16" (8.0mm) and 3/8" (9.5mm). Not all lengths are produced in all diameters. |
| | Generally smaller diameters are sold in shorter lengths only. |

## Electrode Identification

Electrodes are imprinted near the holder end, with a code indicating the character of the electrode wire and its fluxing agent. Electrode codings consist of a letter followed by either four or five numeric digits. The first letter is usually an E, indicating that it is an electrode. The first two numerals in a four digit rated electrode and first three numerals in a five digit rated electrode indicate the minimum tensile strength rating of the welding wire.

*Electrode ID Code*

The strength of the wire is measured in thousands of pounds per square inch. For example, an E-70xx electrode is rated at 70,000 psi tensile strength. An E-110xx electrode is rated at 110,000 psi tensile strength. The digit to the right of the tensile strength rating is either a 1, 2, or 3 and this is used to describe the application of the electrode in terms of position.

- 1 = all positions
- 2 = flat welding and horizontal fillet welds
- 3 = downhand flat

Because the manufacturers of electrodes like to sell as many electrodes as possible, just about every electrode on the market is rated as all position whether it is really suitable for all-position use or not. A good example would be the commonly used E7018 electrode: in sizes under 5/16" (8mm) it is satisfactory for all-position welding, but in the larger sizes it can be difficult to use in anything but the downhand position.

*Electrode Rating*

The final digit, that on the right of the code describes the chemistry of the electrode coating. This is important, probably the most important factor when it comes to the way in which the electrode performs in use. While the welder will not notice the performance difference in electrodes rated at different tensile strengths, the flux coating on each type of electrode can significantly alter performance. It can also determine which polarity options are available. This last digit may be any number from 0 to 8.

*Chemistry of Coating*

# Arc Welding

**Care of Electrodes**

CARE OF ELECTRODES

Electrodes must be protected from damage and deterioration of the coating. Bending the rod can fracture the coating causing it to fall off. Any moisture that is absorbed by the coating will cause an increase in the level of hydrogen in the shielding gas, weakening the weld.

PROTECTION OF ELECTRODES

Electrodes should be stored in a completely dry environment that is sealed and therefore oxygen and moisture free. Once a package of electrodes has been opened, they should be stored in a heated oven at above 212°F / 100°C to keep out moisture. Some electrodes when they become damp can be baked to remove the moisture. Low hydrogen electrodes with an iron constituent in the flux such as the E-7018 category should be discarded when they become damp because the iron starts to oxidized.

**Analysis of Electrode Coatings**

COATING ANALYSIS

Many different chemicals, minerals and ores are used in the production of electrode coatings, but the combinations developed all achieve a few definite purposes. Most coatings contain from 6 to 10 ingredients and the proportions used are determined by the special type of work for which the electrode is to be used.

Electrode coatings fall into the following main groups:

COATING GROUPS

- Arc stabilizers
- Deoxidizers
- Slag formatives
- Shielding gas producers
- Alloy additions
- Metal additions
- Liquid binders

# Function of Ingredients in Electrode Coatings

The function of some of the ingredients used in electrode coatings is described in the following section.

Rutile (Used in, E-6012, E-6013)
A mineral containing approximately 90% titanium oxide. It makes the arc smooth and stable. It forms a hard black slag which gives a smooth finish to the weld.

China Clay, Silica, Mica (Used in, E-6020)
The main function of these minerals is to provide slag volume. Also by varying the additions, the rate of freezing, viscosity and surface tension of the slag can be adjusted.

Potassium Feldspar, Potassium Tianate
These are compounds of minerals containing potassium and are very good as stabilizers and ionizers. Ionization means putting negative ions into the arc atmosphere to facilitate and stabilize the passage of electric current.

Cellulose (Used in, E-6010, E-6011)
Cellulose is a product of wood pulp. It burns in the arc and produces gases which protect the molten metal from oxidation. It forms the inverted cup type shield on the electrode tip, thereby giving direction to the gases and guiding the arc stream.

Ferro-Manganese
(Used in, E-6010, E-6011, E-6012, E-6013)
Ferro-manganese is an alloy containing approximately 80% manganese. It removes oxygen from the arc by combining with it and forming an oxide which passes into the slag.

Iron Oxides (Used in, E-6020, E-6030)
Iron oxides are ores which produce heavy slags capable of holding in solid solution large quantities of oxides which may be formed in the welding operation.

## Arc Welding

**Iron Powder**

Iron Powder (Used in, E-6024, E-6027)
Amounts of iron powder varying from 10 - 50 % are added to electrode coatings. Its purpose is to add additional filler material to that produced by the electrode wire and thus increase the rate of welding. It also improves appearance.

**Water Glass**

Water Glass (Sodium silicate)
(Used in, E-6024, E-6027)
Water glass (sodium silicate) is a heavy liquid, quite viscous and sticky, used for binding the various ingredients of electrode coatings together in a form suitable for extruding them onto the core wire.

## ELECTRODE SELECTION GUIDE

| | | |
|---|---|---|
| E-6010<br>(Cellulose, DP-deep penetration) | DC<br>Reverse Polarity<br>All Position | This shielded arc electrode has a vigorous, pulsating spray type arc. Particularly recommended when quality of deposit, high ductility and tensile strength are essential. For welding root runs on pressure vessels, storage tanks, pipe lines, bridges and all classes of marine work. |
| E-6011<br>(Cellulose, DP-deep penetration) | AC or DC<br>Reverse Polarity<br>All Position | Special coating produces a spray type arc, with strong pulsations which greatly facilitate work in the vertical and overhead positions. Particularly suitable for root runs on pressure piping systems, pressure vessels, building structures, marine work and similar construction where the emphasis is on weld quality. |
| E-6012<br>(Rutile) | AC or DC<br>Straight Polarity<br>All Position | A shielded arc, general purpose electrode for conditions of poor fit-up, single and multiple pass welding. A high burn-off rate and an unusually low spatter loss permits slightly faster deposition at high currents. Deposits are smooth and uniform, of very good appearance, and slag is easily removed. |

## ELECTRODE SELECTION GUIDE

| | | |
|---|---|---|
| E-6013 (Rutile) | AC or DC<br>Straight Polarity<br>All Position | For use on all types of mild steel fabrication. Produces smooth deposit with a minimum of spatter loss. Bead deposit is noticeably flatter and smoother than those of the E-6012 class. Excellent operating characteristics when used with AC transformers having a low open circuit voltage. |
| E-7010-A1 (Cellulose) | DC<br>Reverse Polarity<br>All Position | For producing welds with an exceptionally flat face. Deposited metal quickly solidifies which permits rapid welding in vertical and overhead positions. For use on carbon-moly piping or shapes and castings where molybdenum content does not exceed 0.5%. High tensile strength |
| E-7014 (Iron Powder) | AC or DC<br>Either Polarity<br>Flat and Horizontal | An all-position iron powder electrode of exceptional versatility. Works well on small farm type transformer welders. Very good for vertical down fillet welding. |
| E-7016 (Low Hydrogen) | AC or DC<br>Reverse Polarity<br>All Position | Designed for welding hardened steels susceptible to underbead cracking, cold rolled steels and steels with high sulphur content, weldments which are to be vitreous enameled, and low alloy and mild steels which cannot be stress relieved. |

# ELECTRODE SELECTION GUIDE

| | | |
|---|---|---|
| E-7018 (Low Hydrogen, Iron Powder) | AC or DC<br>Reverse Polarity<br>All Position | For mild, low alloy and "hard to weld" steels. Can be used in place of conventional 7016 or 6016 low hydrogen electrode on all applications. |
| E-7024 (Iron Powder 50%) | AC or DC<br>Either Polarity<br>Flat and Horizontal Position Fillet Welds | Contact type heavily coated electrode for high speed welding, using the drag technique. Deposition rate is exceptionally high because of high percentage of iron powder in coating which mixes with weld deposit thus increasing amount of metal deposited. Slag removal is easy. May be used in applications requiring an E-6012 type of electrode. |
| E11018 (Low Hydrogen, Iron Powder, High Strength) | DC<br>Reverse Polarity<br>All Position | Low hydrogen electrode excellent for welding middle alloy, hardened truck frame channels. |

# ARC WELDING PRACTICE

## Electrode Selection
Electrodes are selected depending on:

- Metal thickness
- Joint preparation
- Position of the joint
- Welding skill

## Amperage
The amperage is determined by the following parameters:

- Size of electrode
- Type of electrode
- Position of weld
- Metal thickness
- Joint type
- Welding skill

## Decimal Equivalent Method
A simple rule of thumb for determining initial amperage settings, is called the "decimal equivalent method." Converting the electrode wire size to its decimal equivalent gives the approximate amperage setting. For example, 1/8" is 0.125", so start with a setting of 125 amps. 1/4" is equivalent to 0.250" so start with a setting of 250 amps. This method is only used by learner welders. The competent welder will pay little attention to the actual current value and adjust the welding current based on the conditions encountered.

## Striking the Arc
There are two methods of striking the arc: scratching and tapping. As the name implies, scratching to start the arc requires dragging the rod lightly across the

base metal, and then moving into position to the area that is to be welded.

Tapping involves remaining closer to the weld area and dropping the electrode directly onto the base metal several times to begin the arc.

## Arc Length

The correct arc length to be observed by the beginner is approximately 1/8" or 3mm. However, this may vary slightly depending on the particular weld and the electrode being used. Too short an arc length may cause an electrode to stick, puddle spatter and blow through. Too long an arc will cause the arc to hiss, electrode spatter and poor penetration.

ARC LENGTH

## Fillet Welds

The most common type of joint that you will be required to weld involves filling an inside corner. This called a fillet weld. Fillet welds can be made in any position, but whenever possible, the piece to be welded is positioned to allow the fastest and strongest weld to be made. This means welding in the downhand position. Fillet welds may be completed in one pass or in multiple passes. This will depend on the size of weld that is required. In multi-pass fillet welds, a planned approach must be taken to as to the position of the welds, and the number of passes made. Whenever a multi-pass weld is made, the previous pass must be thoroughly cleaned so that subsequent passes are not contaminated with slag inclusion.

FILLET WELD

## Root Runs

A root run is the first pass in any butt or fillet weld. In most cases it is required to have excellent penetration. Often special electrodes are used to perform root runs, for instance pressure welding requires that E-6010 or E-6011 DP (deep penetrating) electrodes are used. Though it is commonly not seen, the root run is the most important pass in a multiple pass weld. This root run is especially important in any blind butt weld that must be welded from one side.

ROOT RUN

## Arc Welding

### Welding Bead Terminology

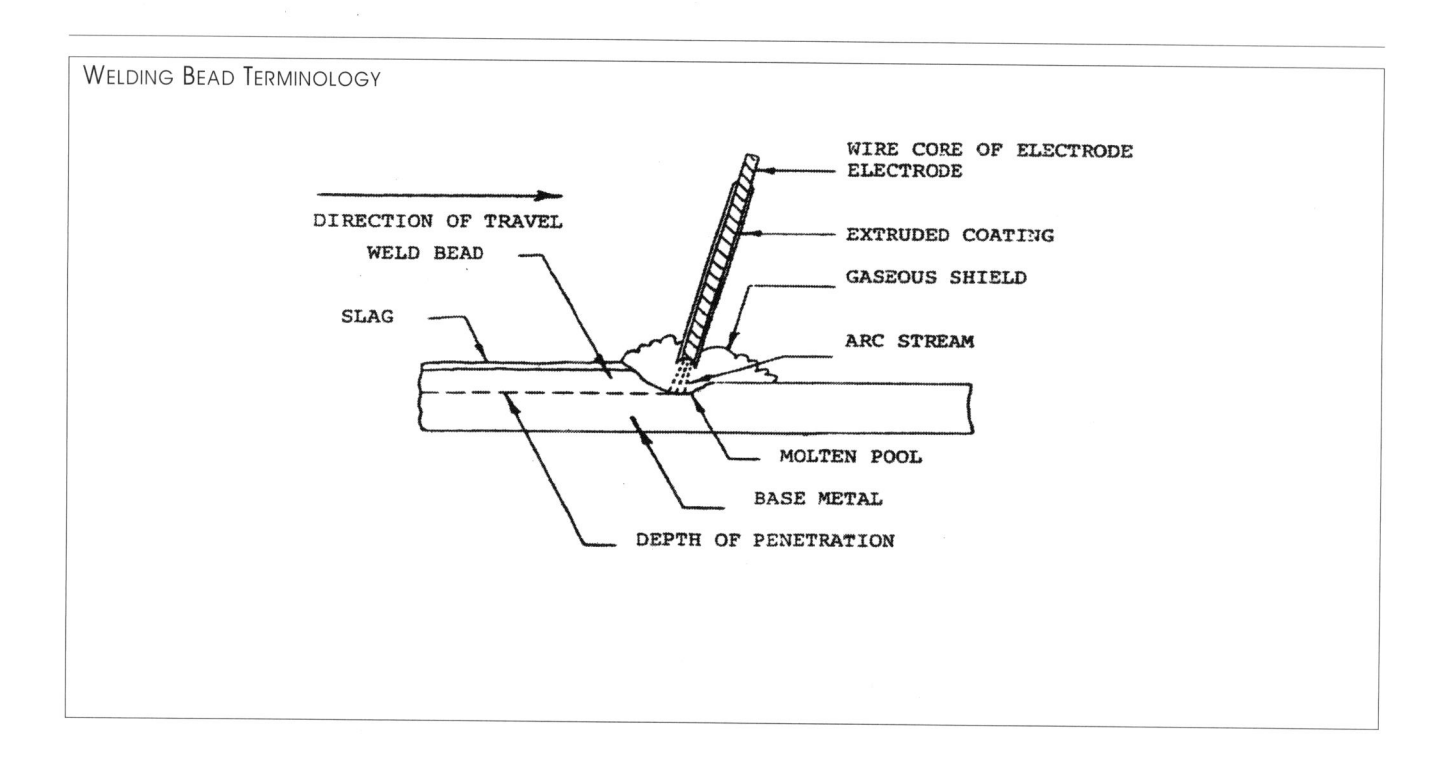

- Direction of Travel
- Weld Bead
- Slag
- Wire Core of Electrode
- Electrode
- Extruded Coating
- Gaseous Shield
- Arc Stream
- Molten Pool
- Base Metal
- Depth of Penetration

### Filet Weld

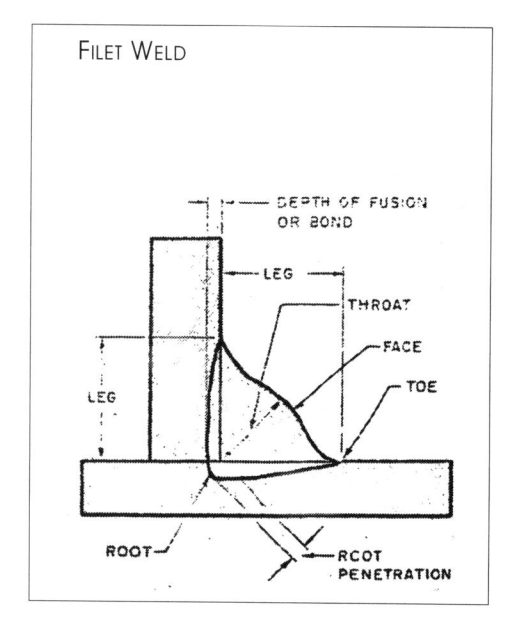

- Depth of Fusion or Bond
- Leg
- Throat
- Face
- Toe
- Leg
- Root
- Root Penetration

### Weaving

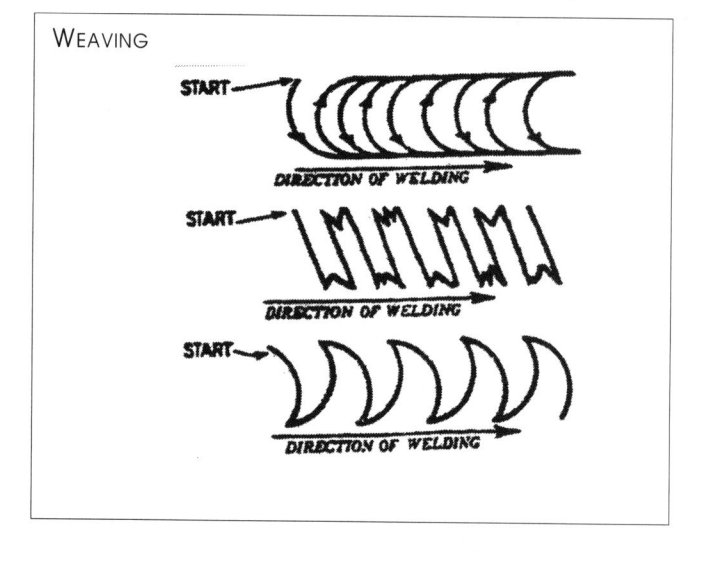

- Start → Direction of Welding
- Start → Direction of Welding
- Start → Direction of Welding

## Weave Passes
A straight bead laid down by the electrode can be called a stringer bead. It is often necessary to lay a bead that is wider than that recommended for the particular electrode being used. This requires a weaving motion of the electrode. In different positions. This weaving motion is achieved in various ways. Capping runs, the final run(s) on a multiple pass weld are performed with a weave technique. Weaving is easy, but like most welding techniques it has to be practiced before it can be perfected.

WEAVE TECHNIQUE

## Lap Welds
This term is used to describe a weld pass joining two pieces of overlapping plate. The technique is essentially that required for a fillet weld. However, the arc angle must accommodate the fact that heat will dissipate rapidly into the lower plate and slowly into the edge plate. Failure to compensate will result in burning away of the edge plate.

LAP TECHNIQUE

## Special Rods and Welding Techniques
For cast iron, and steel alloy welding, special electrodes are available, and require special techniques that may include preheating the metal. Some of these special alloy rods are applied as a base for machining processes. These electrodes are classified as specialty rods and usually have specific requirements that must be met before using them. Always consult the supplier's literature for this information.

SPECIALITY RODS

## Low Hydrogen Electrodes
These are popular electrodes in the automotive repair trade, especially in truck and heavy equipment operations. Low hydrogen electrodes eliminate hydrogen contamination that causes porosity during welding. Low hydrogen electrodes reduce the level of hydrogen by eliminating moisture around the welding arc. They must be kept dry and in fact, exposure to moisture laden air even for a short period will render

LOW HYDROGEN ELECTRODES

them useless. Low hydrogen electrodes require a shorter arc length than that of similar sized electrodes. They should be set at the lowest current they can be run at: this will usually be higher than a comparable non-low hydrogen electrode. Weaving width should be kept to a minimum with low hydrogen rods.

# ELEMENTS OF WELDING PRACTICE

### Beginning to Weld

*Beginning Technique*

When learning to weld always start by running beads, after several attempts you will notice your welds improve. Remember, in arc welding, right handed welders weld left to right and left handed welders weld right to left. A rule of thumb states the width of a bead should be roughly that of 2 electrodes side by side (includes electrode coverings).

### Padding

*Padding Technique*

This is a common way to build up thickness of a metal base. Padding requires a wide weaving motion. Worn areas can be restored this way, and metal can be built up for machining if the correct rod is used. Padding welds must cover the base metal thoroughly and passes must overlap. Cleaning of each pass is necessary to ensure consistent fusion between each run. If several layers of passes are required the direction of the beads can be made at 180 degrees to one another, for subsequent layers.

### Basic Joints and Edge Preparation

*Joints and Edge Preparation*

Before performing any type of weld you must always have a:

- Clean metal surface
- Good fit
- Edge prepared
- Proper tacks

## Positional Welding

Besides downhand passes some other welding positions you might have to deal with are:

➢ Vertical

➢ Overhead

➢ Horizontal

Each has its own technique and, despite the AWS coding, some electrodes are unsuitable for certain types of positional welding.

Welding Positions

# METAL INERT GAS (MIG) WELDING        5

Metal Inert Gas or MIG welding, as it is usually known, is a semi-automatic process which can performed manually or in robotic machining applications. Many automobile, truck and heavy equipment manufacturing assembly lines use automated MIG welding due to its good appearance and low-labor clean-up. In service garages, MIG has almost entirely replaced arc welding for the following reasons:

➢ Less skill required to operate the equipment.

➢ Much faster weld deposit rate.

➢ Minimal post weld clean-up and no slag removal required.

➢ High current, low voltage reduces the overall heat affected area of the base metal.

➢ No hydrogen inclusion in the weld affected area.

MIG welding can be used to weld most metals including low and middle alloy steels, stainless and heat resistant steels, copper, bronze and aluminum and its alloys.

## MIG PROCESS

All MIG processes use a flexible, consumable electrode of bare wire that is driven through a spool and fed to either a manually or automatically operated gun. The consumable wire electrode is driven through an outer flexible cable by motor-driven rollers whose speed may be adjusted.

# JOINT AND WELD CLASSIFICATIONS / TYPES OF JOINTS

THERE ARE ONLY FIVE BASIC TYPES OF JOINTS. THEY CAN, HOWEVER, BE USED IN COMBINATIONS.

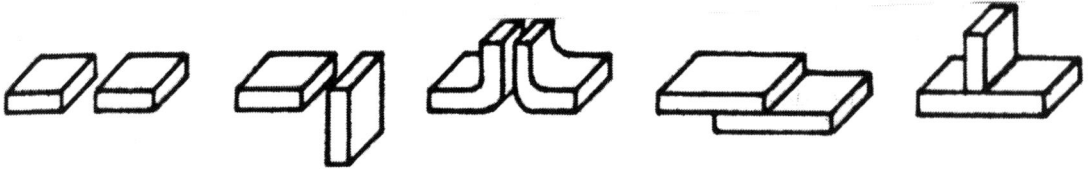

- B - Butt
- C - Corner
- E - Edge
- L - Lap
- T - Tee

## TYPES OF WELDS

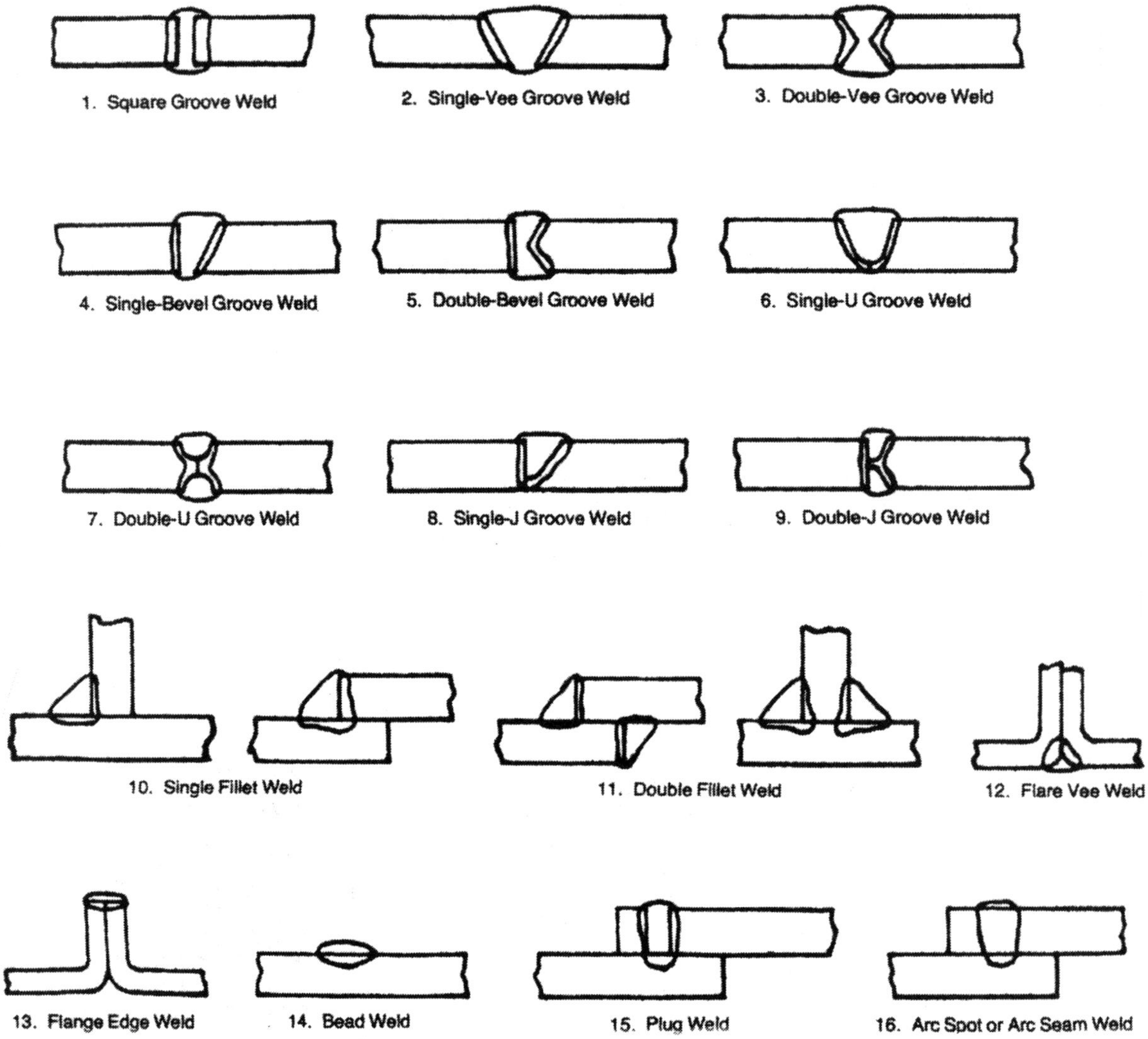

1. Square Groove Weld
2. Single-Vee Groove Weld
3. Double-Vee Groove Weld
4. Single-Bevel Groove Weld
5. Double-Bevel Groove Weld
6. Single-U Groove Weld
7. Double-U Groove Weld
8. Single-J Groove Weld
9. Double-J Groove Weld
10. Single Fillet Weld
11. Double Fillet Weld
12. Flare Vee Weld
13. Flange Edge Weld
14. Bead Weld
15. Plug Weld
16. Arc Spot or Arc Seam Weld

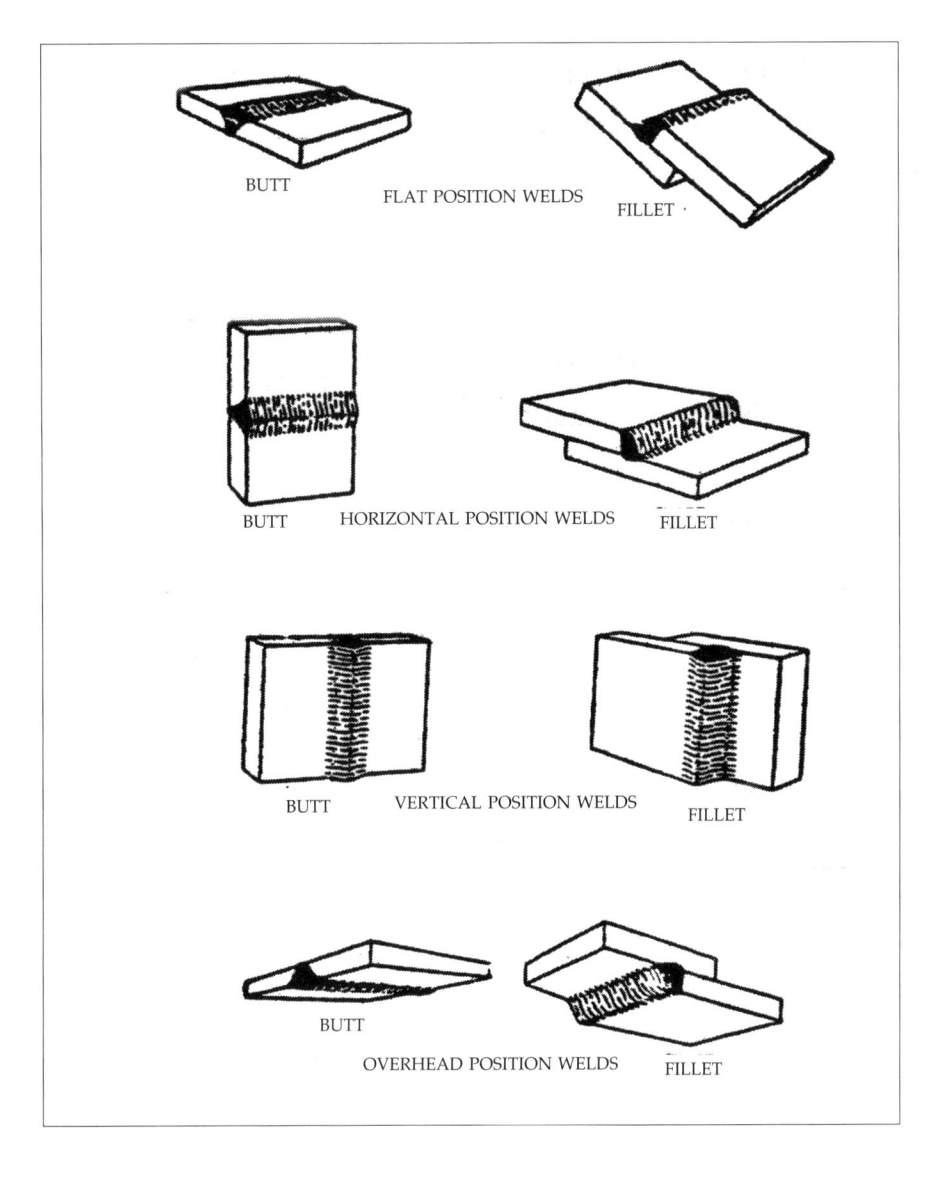

MIG WELDING

FLAT POSITION WELDS

BUTT   FLAT POSITION WELDS   FILLET

HORIZONTAL POSITION WELDS

BUTT   HORIZONTAL POSITION WELDS   FILLET

VERTICAL POSITION WELDS

BUTT   VERTICAL POSITION WELDS   FILLET

OVERHEAD POSITION WELDS

BUTT   OVERHEAD POSITION WELDS   FILLET

BUTT WELDING TERMINOLOGY

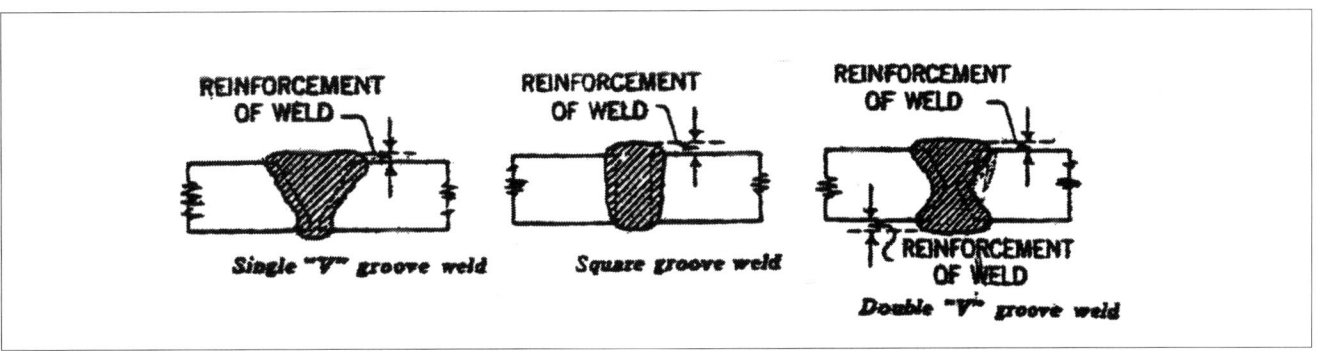

REINFORCEMENT OF WELD

Single "V" groove weld

REINFORCEMENT OF WELD

Square groove weld

REINFORCEMENT OF WELD

REINFORCEMENT OF WELD

Double "V" groove weld

# MIG Welding

**MIG Process**

## POSITION WELDING

The burn-off rate of the wire electrode must be balanced by the rate of wire feed. This means that wire feed rate to some extent determines the current setting used. As the wire exits the MIG gun, an arc is formed and it melts to a molten puddle formed by the heat of the arc and the base metal. The MIG welding process requires that an inert gas be fed to the gun to shroud the arc and molten weld pool. The actual gas used depends on the process.

The inert gas is routed from a storage cylinder, through the flexible cable and out to the gun assembly. A cone in the gun helps to direct the shielding gas and to contain it around the immediate weld area. Heavy duty or extended duty cycle MIG welding units may be water-cooled. The cooling water is pumped from the welding station, through the flexible cable and gun, and then returned to the station.

**Shielding Gas Functions**

## SHIELDING GASES

The shielding gas used in MIG welding has 2 functions:
- Protects the arc and molten metal from atmospheric contamination.
- Influences the arc condition.

**Shielding Gases**

The gases used are:
- Argon
- Helium
- Carbon dioxide

These gases are often mixed to improve the arc condition or the finished appearance of a weld. For instance, when argon is used to weld aluminum, it is often mixed with 5% oxygen: the oxygen oxidizes the finished weld and gives it a shiny appearance.

# WELDING EQUIPMENT

The equipment used for MIG welding must be able to perform properly over the range of voltages and currents required for the process.

## Power Supply Unit
Power supply units used produce a flat DC voltage and may be rated for continuous duty cycle or by percentage duty cycle. Most MIG power supply units are equipped with an ammeter and a voltmeter. Provision is made for fine and coarse voltage adjustment.

## Wire Feed Units
The wire feed unit consists of a governed electric motor and a drive roll assembly that feeds the wire along the flexible hose that connects the power control unit to the gun. The motor speed is adjusted to give the desired current. Typically, the motor speed window extends 20% above and 20 % below the set range.

## Welding Guns
Welding guns used in MIG welding can be gas cooled or water cooled. In a gas cooled gun, the inert shielding gas performs the function of cooling the gun. In applications requiring high current, high duty cycle, water is used to cool the gun. Most welding guns are balanced to the hand. This enables them to be used for prolonged periods without causing operator fatigue.

## Welding Wire
The welding wire used in MIG welding processes is sold in spools. The spools connect to the wire feed unit so that the wire can be driven through the flexible cable and gun during welding. The wire is made either out of the same metal as the base metal to be welded, or an alloy that is designed to improve the metallurgical characteristics of the weld. Mild steel wire on spools is generally coated with corrosion inhibitor. This is usually copper colored and prevents oxidization of the wire before welding.

# MIG Welding

Courtesy of Air Products Canada Ltd.

**WELDING GUN**

Welding guns can be gas cooled or water cooled (for high currents). The gun is balanced about the hand grip to reduce operator fatigue and to aid manipulation.

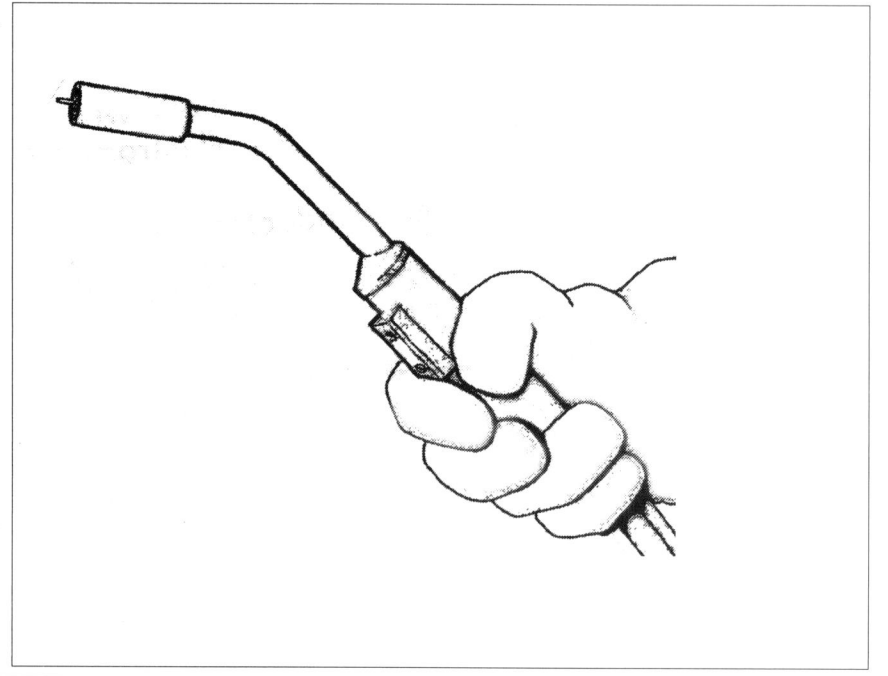

**POWER SUPPLY UNIT**

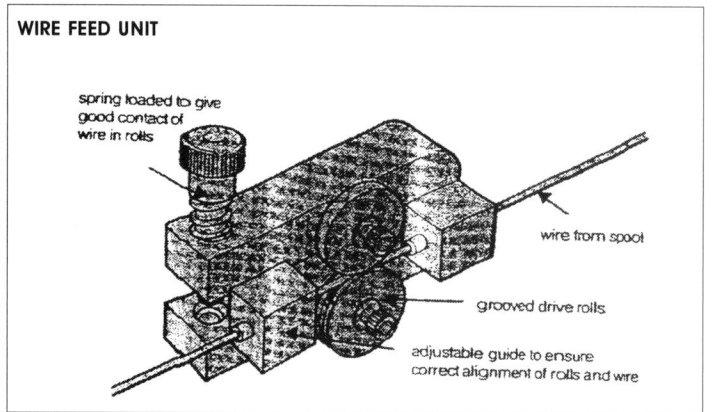

**WIRE FEED UNIT**

- spring loaded to give good contact of wire in rolls
- wire from spool
- grooved drive rolls
- adjustable guide to ensure correct alignment of rolls and wire

Wire feed units include a governed motor and a drive roll assembly which feeds the wire along a flexible hose to the gun. The motor speed is adjusted to give the desired current. Ideally, the motor speed extends to 20% above and 20% below the anticipated range.

MIG WELDING

Courtesy of Air Products Canada Ltd.

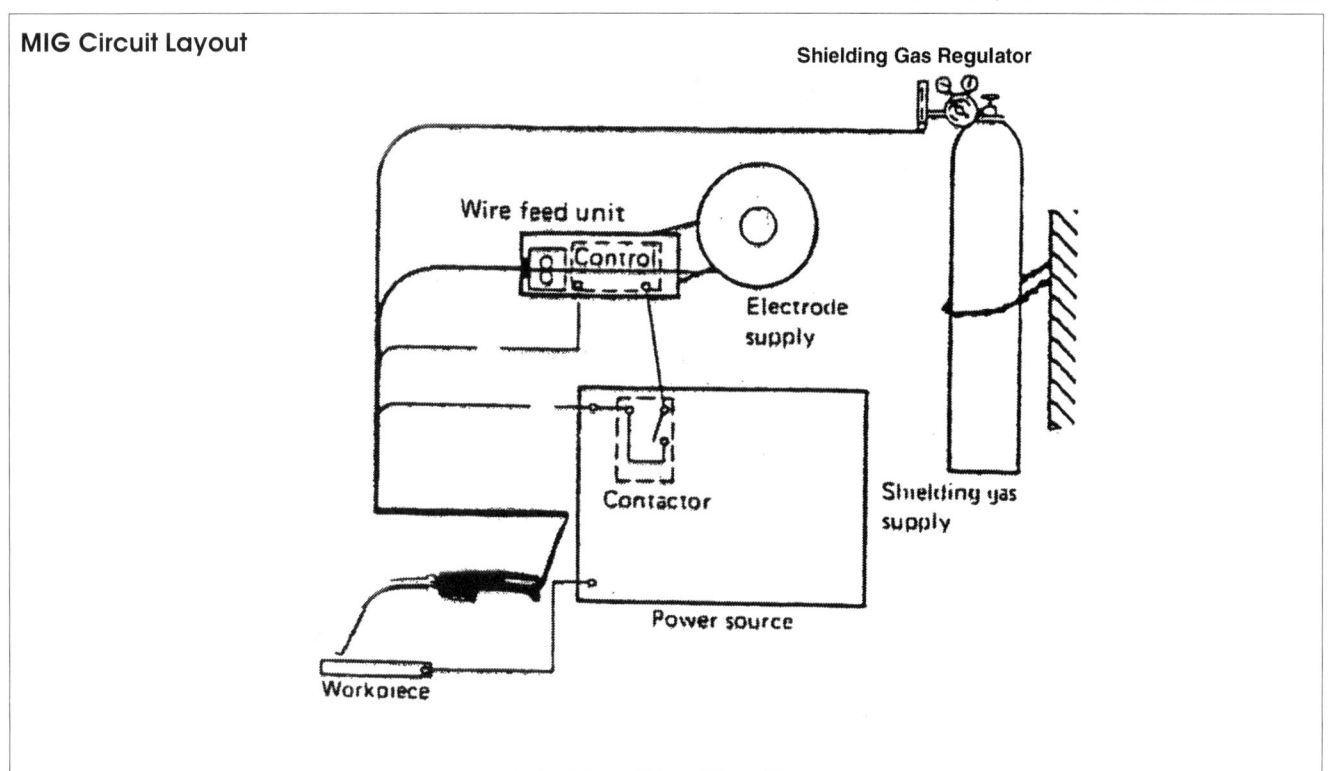

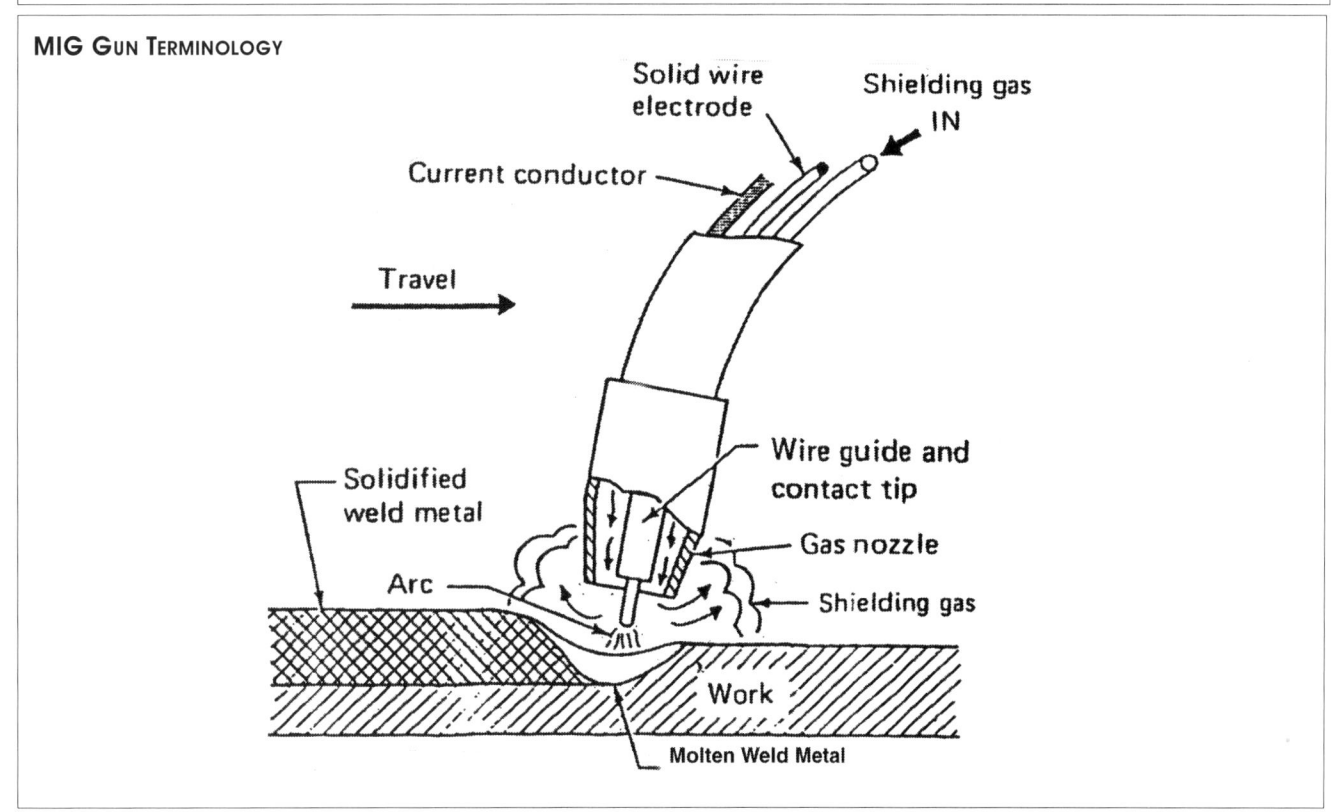

## MIG Circuit Layout

## MIG Gun Terminology

MIG Welding

## SHORT CIRCUIT TRANSFER OPERATION

Short circuit transfer operation enables satisfactory welds to be deposited at currents below 180A. With this technique, the end of the electrode touches the surance of the weld pool 50 to 100 times each second to transfer metal to the weld pool.

The output from the power supply unit is thus periodically short circuited and an inductance is added to the welding circuit to control the variations in current. Between each short circuit the arc heats the weld pool.

Courtesy of Air Products Canada Ltd.

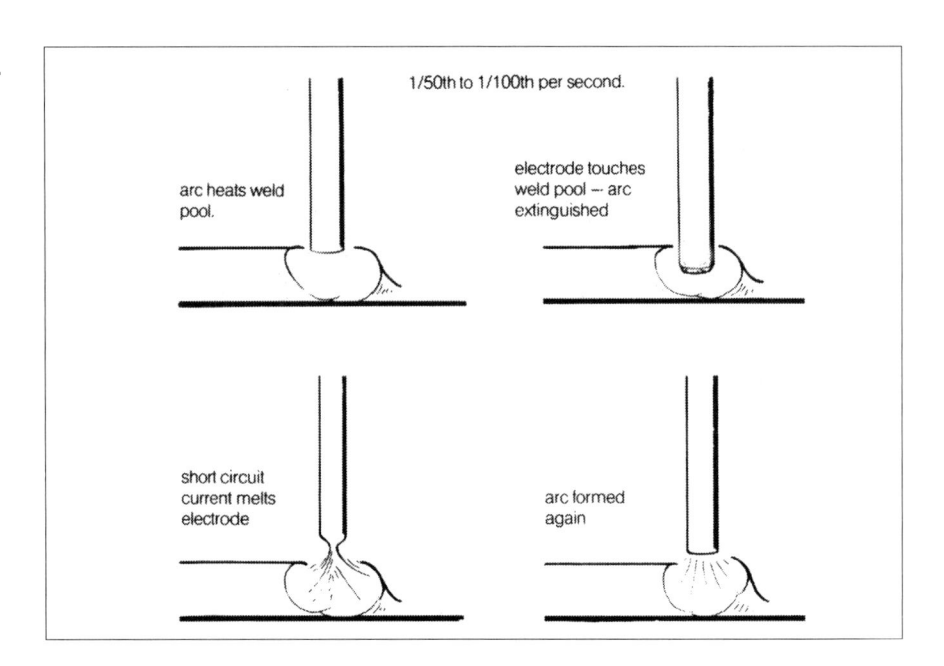

### ELECTRICAL STICKOUT FOR SHORT CIRCUIT TRANSFER MODE
The contact tip should be flush with the end of the nozzle or extend a maximum of 1/8" (3.2mm) as shown.

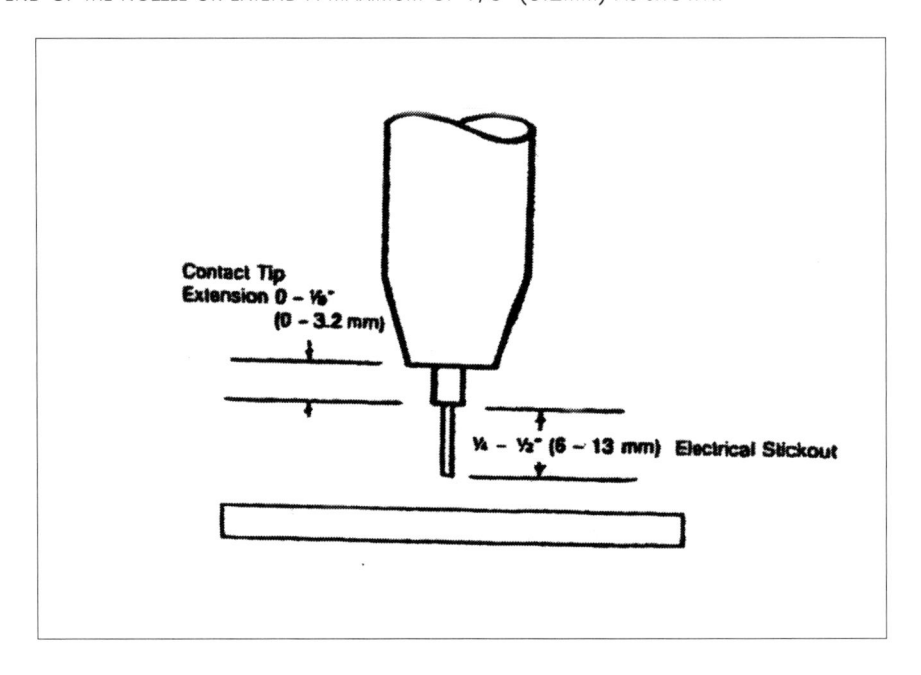

# MIG WELDING TECHNIQUES

There are two basic methods of operating MIG welding, each dependent on exactly what is being welded. Some MIG welding equipment is limited to the first method, which we will describe as short circuit: it can also be known as dip transfer. The other method, spray transfer, requires higher operating currents and higher gas densities. The term pulse transfer is sometimes used to describe a procedure that is midway between short circuit and spray transfer. All three methods are described here.

MIG Welding Methods

## Short Circuit Transfer

Lower arc voltage and current/ wire speed settings are required for short circuit transfer. Most low cost MIG welding stations are designed to operate only on short circuit method. Molten wire transfer take place in globules of molten metal, intermittent with short circuiting of the arc. When the consumable wire from the MIG gun contacts the pool, there is a momentary rise in current which has to be sufficient to melt the tip of the wire into a globule which is then sucked into the molten puddle in the base metal by surface tension. As the molten globule separates, the arc is re-established and gradually decreases in length as the wire feed rate gains on the burn-off until short circuiting recurs. The arc/ short circuiting cycle occurs from 50 to 100 times per second.

Short Transfer/Dip Transfer

Short circuit transfer permits satisfactory welds to be deposited at currents below 180 amps. Wire electrode diameter is typically 0.9 mm or 1.2 mm but smaller sizes can be used in the lower range current settings. In most cases, an electrode wire diameter should be selected that allows the drive motor to operate in the middle one third of its speed range window.

# MIG Welding

## WIRE FEED SPEED TO CURRENT CHART

| Short Circuit Electrode Wire Diameter | Current Range in Amperes | Feed Speed in mm/min |
|---|---|---|
| 0.6 mm | 60 - 100 | 5.8 - 10.4 |
| 0.8 mm | 70 - 150 | 3.8 - 8.0 |
| 0.9 mm | 80 - 170 | 3.0 - 5.6 |
| 1.2 mm | 100 - 180 | 2.2 - 4.0 |

### Spray Transfer

In normal arc or "stick" welding, the metal in the electrode wire or rod is transferred in molten globules from the electrode to the base metal being welded. In MIG spray transfer, the current fed to the wire electrode is increased along with the wire feed rate. This results in the molten droplets becoming smaller and the transfer occurs in the form of a fine spray.

The shielding gas used has a great influence on the minimum current for the spray transfer. For instance, much greater current values are required when using $CO_2$ than when using argon to obtain an equal droplet rate. When spray transfer is being performed, the arc is never extinguished. This results in high arc energy output, deep penetration, considerable puddle dilution and high deposition rates. The method is especially suitable for production welding on thicker metals in the flat or downhand position.

If currents become excessively high, there is a danger of entrapping oxides in the weld metal. The high voltage drop across the arc plus the high current density in the wire electrode make the process much better suited to the welding of heavy metal sections. The high currents produce a very directional arc and strong magnetic fields. When argon is used as a shielding gas, the forces acting on the droplets are well balanced so that they move slowly from wire to work with almost no spatter. With $CO_2$, the forces acting on the droplets are less even and there is a tendency to produce spatter. Today, special-mix gases are generally used. These special-mix gases are formulated exactly to the welding procedure to produce the best results.

# MIG WELDING

## SPRAY TRANSFER PRINCIPLES

Courtesy of Air Products Canada Ltd.

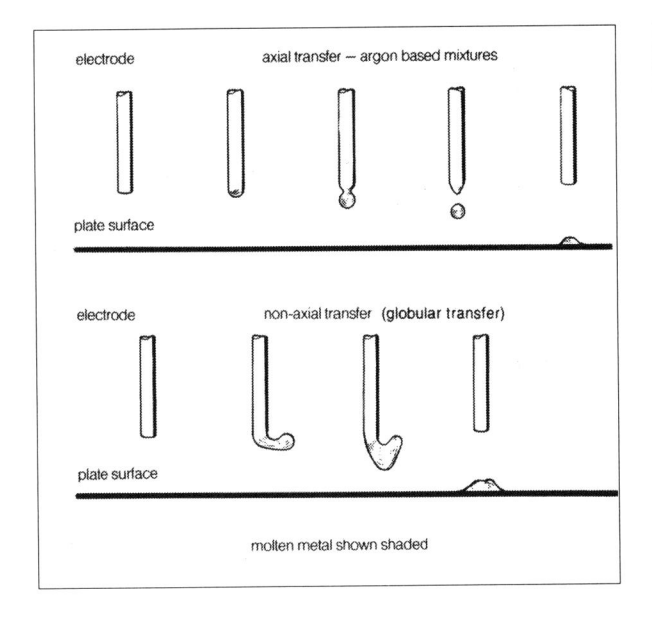

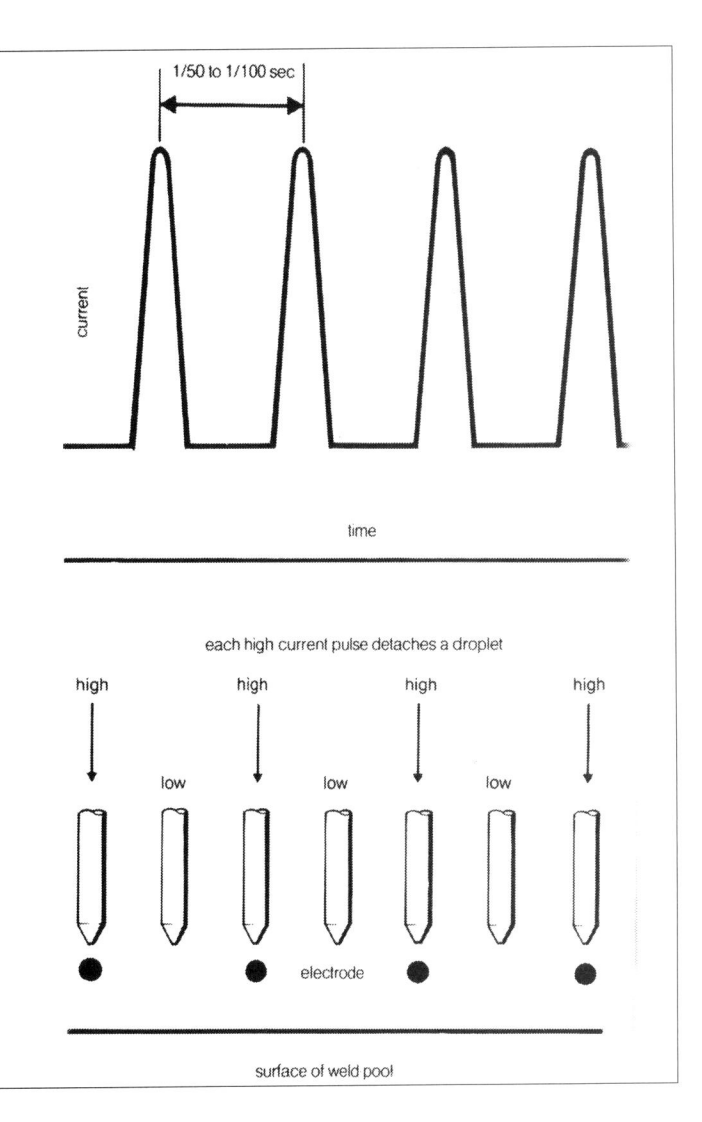

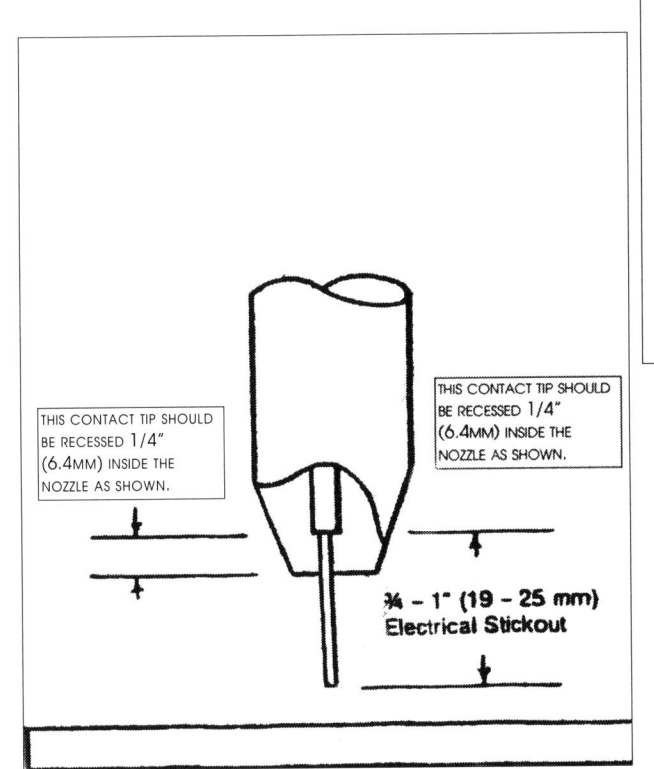

electrode — axial transfer — argon based mixtures

plate surface

electrode — non-axial transfer (globular transfer)

plate surface

molten metal shown shaded

1/50 to 1/100 sec

current

time

each high current pulse detaches a droplet

high — low — high — low — high — low — high

electrode

surface of weld pool

THIS CONTACT TIP SHOULD BE RECESSED 1/4" (6.4MM) INSIDE THE NOZZLE AS SHOWN.

THIS CONTACT TIP SHOULD BE RECESSED 1/4" (6.4MM) INSIDE THE NOZZLE AS SHOWN.

¾ – 1" (19 – 25 mm) Electrical Stickout

## ELECTRICAL STICKOUT FOR SPRAY TRANSFER MODE

# MIG Welding

**Pulse Transfer**

PULSE TRANSFER

This term is used to describe MIG welding using settings that are midway between short circuit and spray transfer. The idea is that pulse transfer offers the advantages of both other methods. The arc is maintained by a low level background current while a pulse of high level current is injected through the arc. Each "pulse" results in a droplet detaching itself from the electrode wire and being transferred into the weld puddle. The pulse frequency depends on the equipment type being used, but usually falls into a 50 to 100 times per second window.

CRITICAL WELDING PARAMETERS

The critical parameters in MIG pulse welding are:
- Background current
- Pulse height (peak current)
- Pulse frequency (pulses per second)
- Pulse duration (pulse length)
- Mean current (related to wire feed speed: defines heat input)

All the above parameters must be correctly set to produce good pulse transfer results.

STANDARD/METRIC WIRE SIZES

Standard to Metric Comparison of Common Wire Sizes
Most manufacturers of MIG welding wire market their product using metric gauge values. The following chart converts Standard and Metric sizes. It should be noted that 1 inch is equivalent to 25.40 millimeters or 2.54 centimeters.

| METRIC | STANDARD |
|---|---|
| 0.6 mm | 0.024 inches |
| 0.8 mm | 0.031 inches |
| 0.9 mm | 0.035 inches |
| 1.2 mm | 0.047 inches |
| 1.6 mm | 0.063 inches |
| 2.0 mm | 0.079 inches |

# MIG Welding

Courtesy of Air Products Canada Ltd.

## Electrode Angles

Usually in metal arc gas shielded welding the electrode is pointed in the direction of travel (i.e. forehand). This gives the best penetration and assists weld pool control. Use of correct angles ensures good fusion of the parent plate.

When a surface layer is being deposited, it is often necessary to limit the amount of parent metal melted into the weld pool. In these cases the electrode can be pointed towards the completed weld, i.e. backhand.

Note: the travel speed is critical with backhand techniques if lack of fusion is to be avoided.

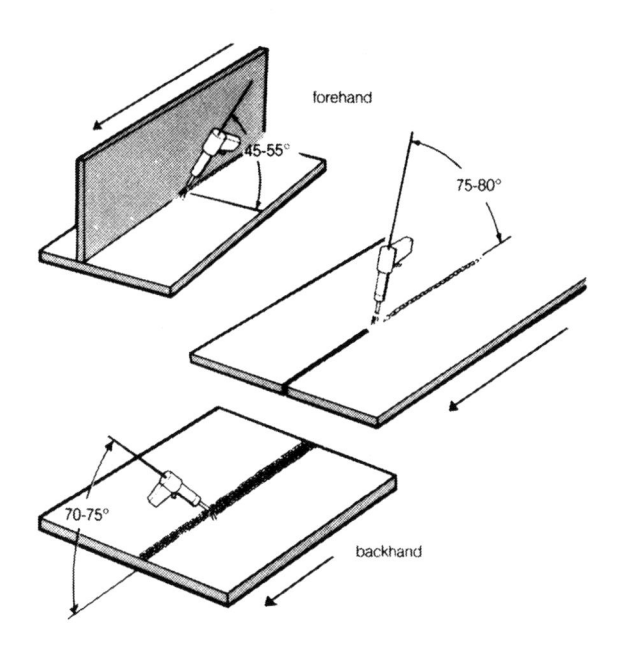

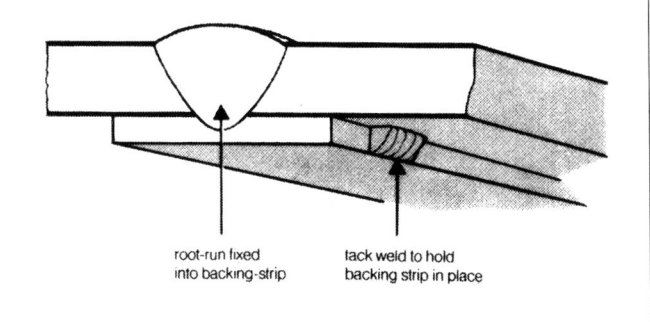

**Backing Strip:** A length of metal of the same composition as the parent strip; the strip is usually left in place after welding.

## Spray Techniques — Edge Preparations

| Type | Thickness | Low Carbon steel and stainless steel | Aluminium |
|---|---|---|---|
| Square edge | 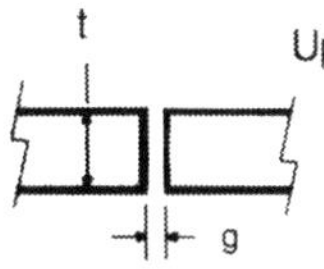 Up to 6mm | $g = \frac{1}{2}t$ | $g = \frac{1}{2}t$ |
| Single V | 6mm to 18mm | $A = 60°$<br>$Rf = 1.5mm$ max<br>$g = 1mm$ max | $A = 65.75°$<br>$Rf = 1.5$ max<br>$g = 1.5mm$ |
| Double V | Above 18mm | $A = 50°$<br>$Rf = 1$ to $2mm$<br>$g = 1.0$ max | $A = 80-90°$<br>$Rf = 1.5$ max |

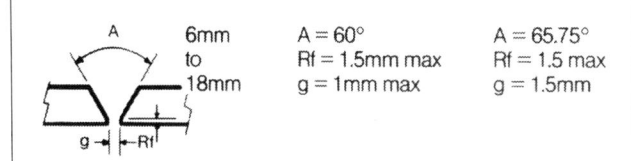

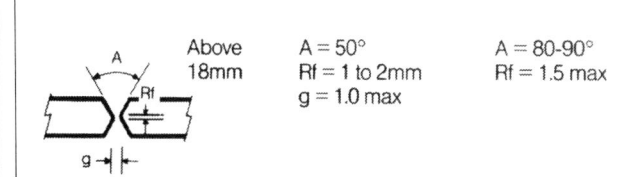

**Backing bars and strips:** The deep penetration characteristic of spray transfer makes it difficult to weld root runs without producing burn-through. The molten underside of the weld metal must be supported. This can be achieved by using: **Backing bar:** A copper bar with a groove; the bar is removed after welding.

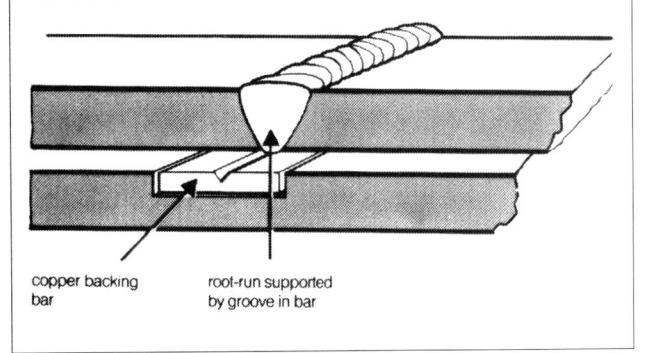

# MIG Welding

Courtesy of Air Products Canada Ltd.

**Joint Preparation:** Edges must be cut square without burrs. Grease, paint and scale must be removed before assembly.

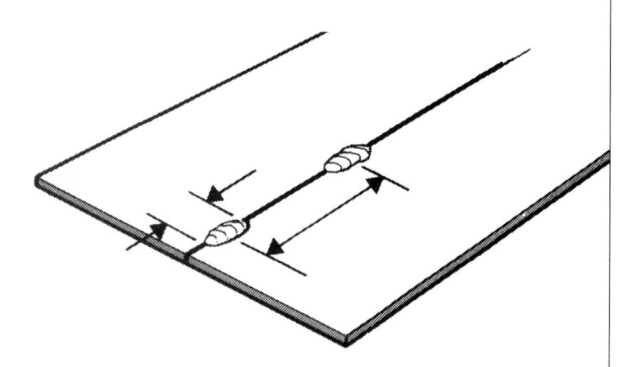

Sheets must be held in alignment by clamps or by tack welds at about 50mm intervals.
Gap allowed between sheet edges depends on joint type and thickness.

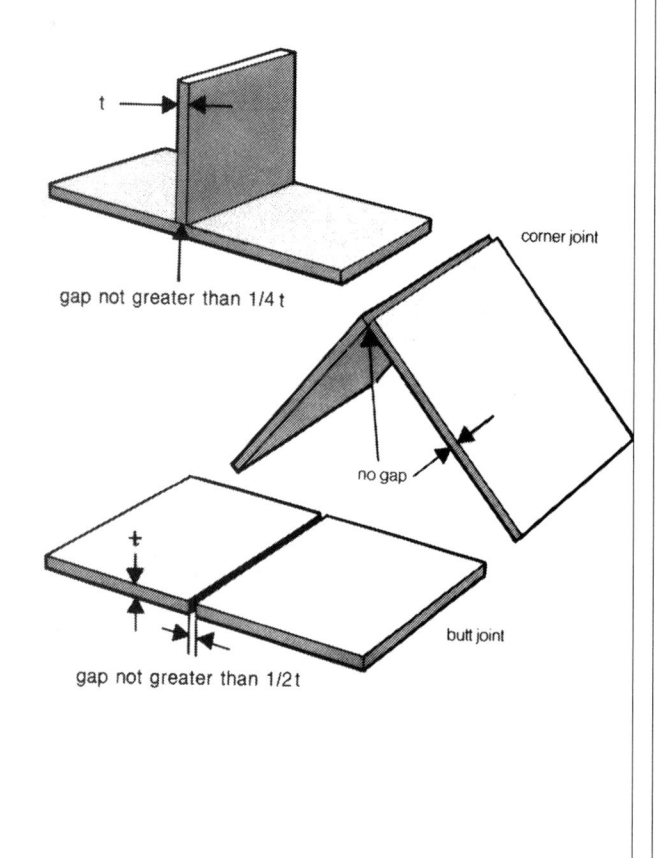

gap not greater than 1/4 t

corner joint
no gap

butt joint
gap not greater than 1/2 t

**Welding Procedure:** For joints welded in the flat, horizontal-vertical and overhead position, the electrode is pointed in the direction of welding at an angle of **80°** to **75°** to the sheet surface.

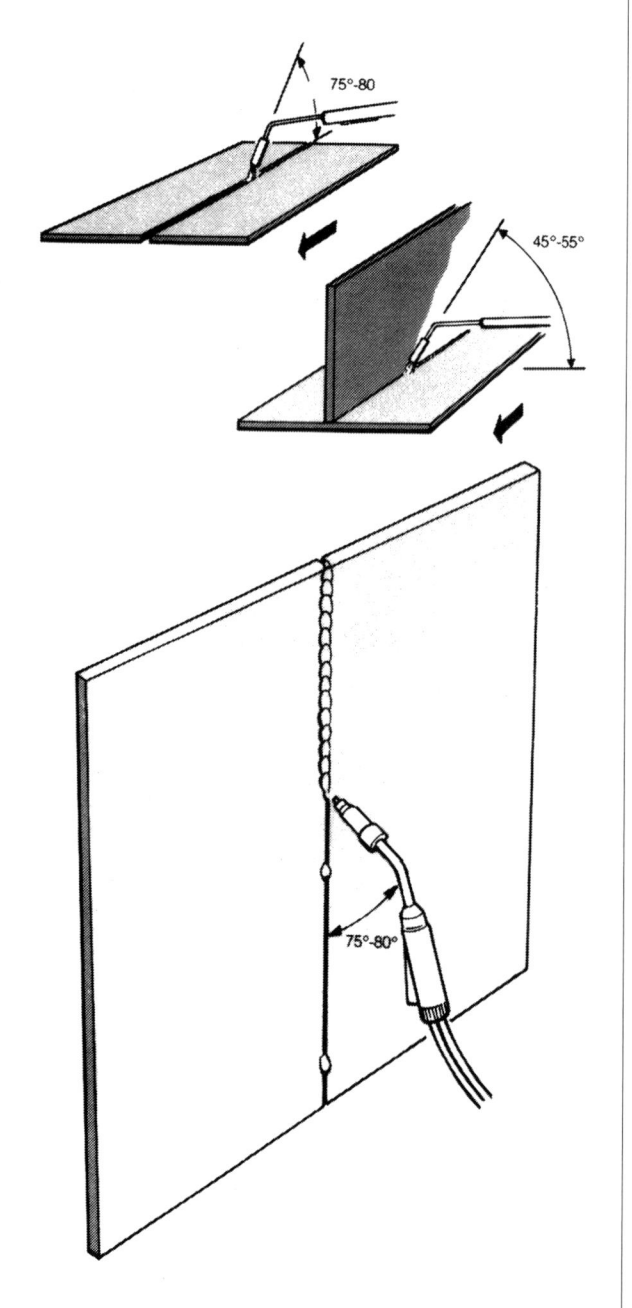

75°-80°

45°-55°

75°-80°

For joints welded in the vertical position, the gun is moved down the joint, with the electrode pointing upwards.

# MIG Welding

## Welding Procedures

Courtesy of Air Products Canada Ltd.

Voltage — high / correct / low

nozzle to plate distance — kept at about 19-25mm by welder

arc length/Tip to Work

Positional welding. Short arc or pulse transfer.

current too low

current too high

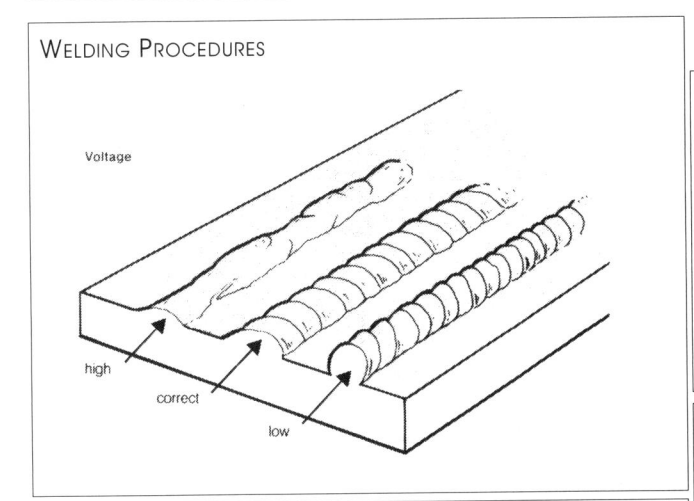

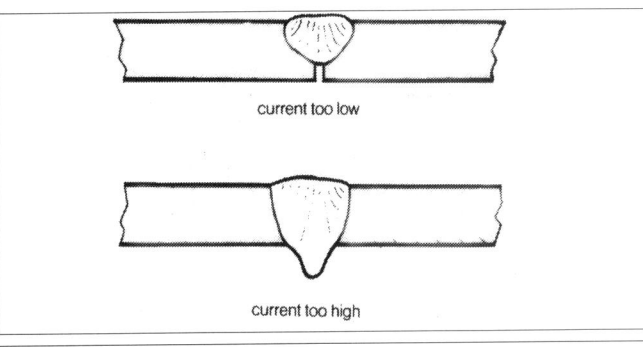

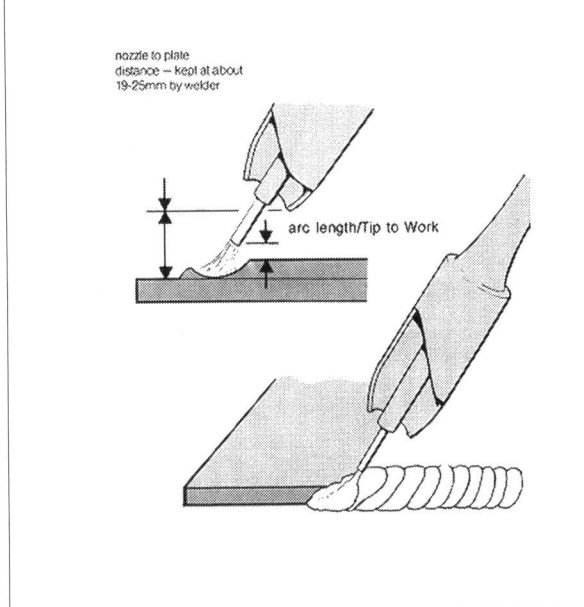

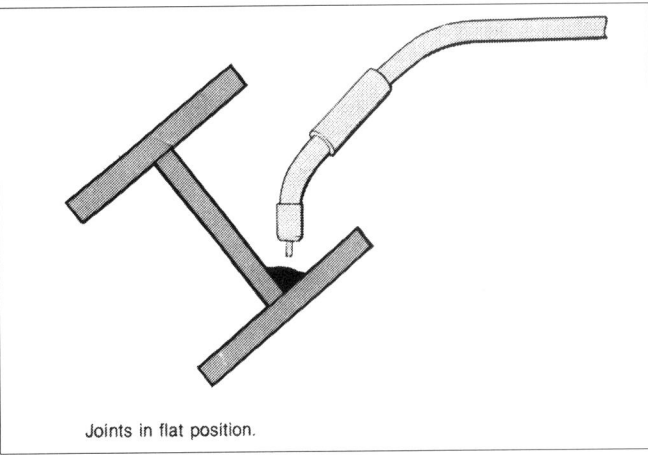

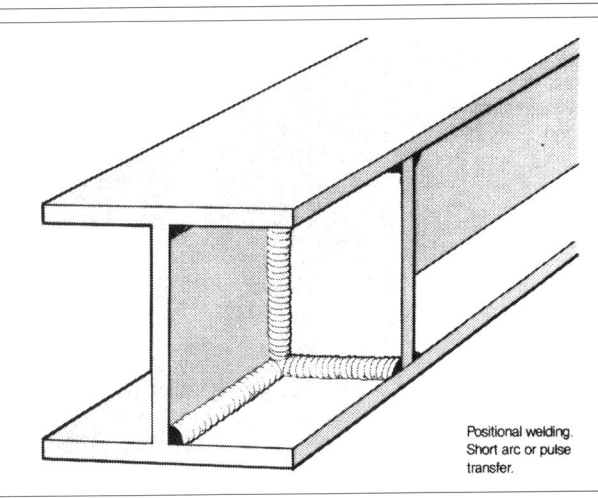

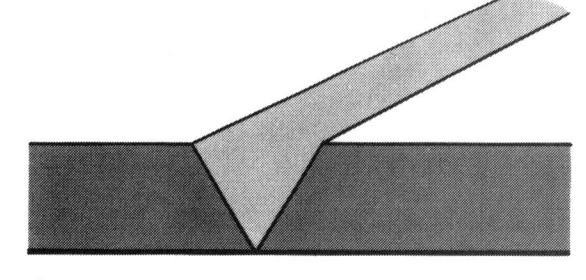

Joints in flat position.

Single-V: normally for thicknesses between 6mm and 15mm but can be used for greater thicknesses where access is from one side only.

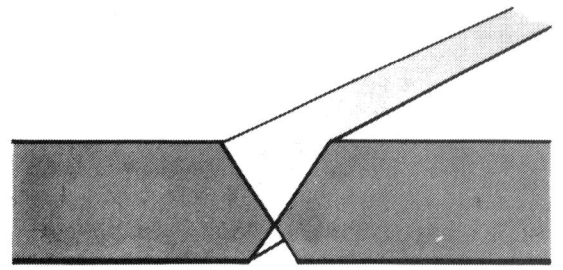

Double-V: for thicknesses greater than 15mm. The joint can be symmetrical, but better flatness after welding can be achieved with an assymetrical preparation.

# MIG Welding

## Typical Welding Defects
*Courtesy of Air Products Canada Ltd.*

An acceptable weld should have the following characteristics:
A  Smooth surface; uniform ripples
B  No undercut at toes
C  Smooth transition from weld surface to parent metal surface
D  No internal flaws
E  No discontinuities at the fusion boundary
F  Uniform penetration bead
G  Good root fusion

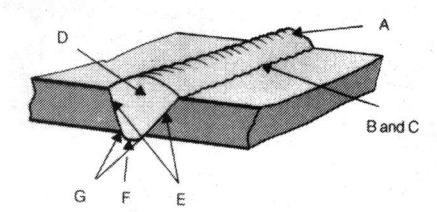

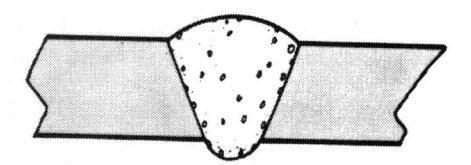

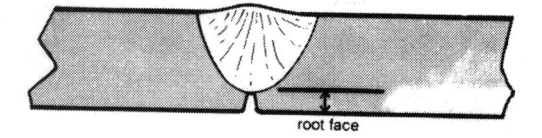

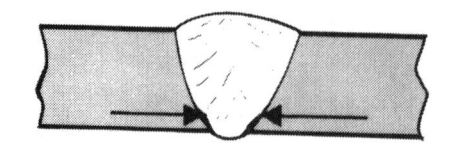

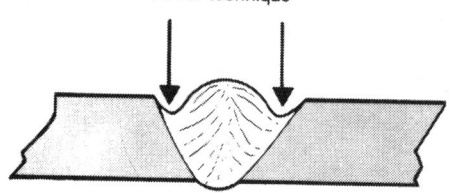

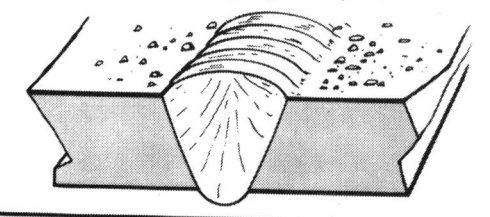

| Defect | Probable causes |
|---|---|
| **Porosity** | 1 Gas flow too low or too high<br>2 Blocked nozzle<br>3 Leaking gas lines<br>4 Draughty conditions<br>5 Nozzle/work distance too long<br>6 Painted, primed, wet or oily work<br>Wet or rusty wire |
| **Lack of penetration** | 1 Current too low<br>2 Preparation<br>3 Root face to thick<br>4 Root gap too small<br>5 Worn contact tip causing irregular arcing<br>6 Irregular wire feed<br>7 Poor technique<br>8 Mismatched joint |
| **Lack of fusion** | 1 Voltage too low<br>2 Current too low or too high<br>3 Wrong torch angle<br>4 Incorrect inductance setting |
| **Undercut** | 1 Speed too high<br>2 Incorrect current/voltage<br>3 Poor technique |
| **Spatter** | 1 Incorrect gas<br>2 Voltage too low/amp too high<br>3 Rusty or primed plate |
| **Centre line crack** | 1 Depth of weld run greater than 1.5 x width of vee at surface<br>2 Low voltage, high current<br>3 High sulphur in steel<br>4 Incorrect electrode composition/ stainless steel and aluminium alloys |

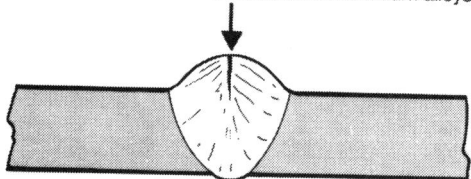

# TUNGSTEN INERT GAS (TIG) WELDING

# 6

## TIG Welding

TIG or tungsten inert gas welding is not commonly used in automotive service facilities. It is more likely to be found in racecar chassis builder operations and larger truck and heavy-duty garages. TIG welding is a shielded gas process that differs from MIG welding primarily in that the electrode is non-consumable. In terms of technique, manual TIG welding relates more to oxy-acetylene welding than MIG or arc processes. Perhaps TIG welding is most noticeable for the kind of picture-perfect finish that can be produced by a good practitioner and if you are in the habit of admiring a fine weld on a vehicle, chances are, this has been courtesy of a TIG welder. It can be used on most metals but because it is slow to set up and execute, it tends to be reserved for stainless steels, aluminum alloys and special metals. TIG can also be performed automatically or robotically in manufacturing, but those procedures are beyond the scope of this book.

## Inert Shielding Gas

In both the MIG and TIG processes, the inert shielding gas used shrouds the molten weld area much better than the flux coating of an arc electrode or the flame of an oxy-acetylene torch. The shielding gas is routed to the TIG torch by means of a dedicated hose. A tungsten non-consumable electrode is used to heat the base metal to a molten state. Filler wire is then dipped into the weld puddle by the welder using a technique similar to the oxy-acetylene welder. Many TIG torches are water-cooled using dedicated tubing within the torch hose assembly. An advantage of TIG welding is the precise view of the molten weld puddle that the welder is given, permitting adjustment to such factors as contamination and irregularities.

# TIG Welding

TIG welding permits high strength welds with the minimum filler deposition, a real advantage where weight is important such as in a racecar chassis. The biggest disadvantage is the time required to prepare a TIG welding station and the slowness of performing the weld itself compared with MIG.

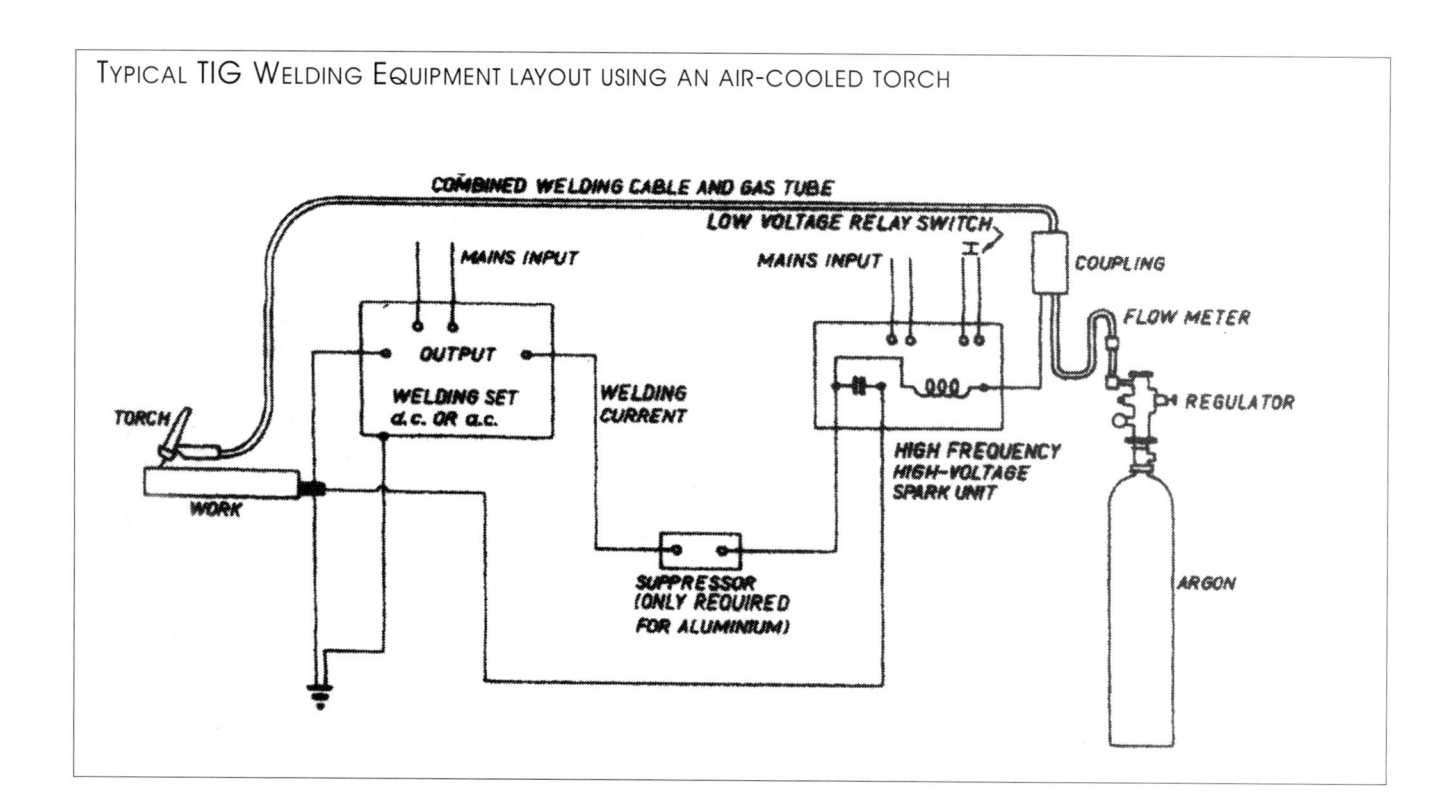

**Typical TIG Welding Equipment layout using an air-cooled torch**

## TIG PRINCIPLE OF OPERATION

Operating Principle of TIG

In TIG welding, the arc burns between a tungsten electrode in the welding torch and base metal to be welded within a shield of inert gas. The shielding gas excludes atmospheric air from contaminating the molten weld puddle. Unlike the electrode in manual arc and MIG welding, the tungsten electrode does not transfer to the weld puddle and when used properly, vaporizes slowly during welding. Tungsten electrodes have small amounts of other elements added to improve the welds that are produced and because of this are often color coded to each specific procedure.

# TIG WELDING

For metals having refractory surface oxides such as aluminum, aluminum alloys, magnesium alloys and aluminum bronze, AC current is required. This category of metals tends to be difficult to weld because the oxides formed on the skin have melting temperatures that exceed that of the base metal or alloy. When oxy-acetylene welding is used to weld these metals, use of corrosive fluxes and a high level of operator competency is required.

REFRACTORY SURFACE OXIDES

Most other metals are TIG welded using DC current. Carbon steel, middle to high alloy steels, copper, stainless steels, nickel, silver, and titanium can all be welded using TIG and DC current.

## TIG EQUIPMENT

A TIG welding station set up for manual welding uses the following equipment:

TIG WELDING STATION

### TIG Power Source
The TIG power source is a transformer/rectifier unit capable of producing both AC and DC current characteristics. The power source should be equipped with a high frequency unit capable of initiating a spark from the tungsten electrode to the base metal: this is required to avoid contacting the work with the tungsten electrode.

TIG POWER SOURCE

The inert gas used in a TIG welding station is usually routed through the transformer/rectifier unit for purposes of controlling its flow during welding cycles. Many TIG guns are water-cooled. The water cooling medium is held in a tank integral with the power supply station and pumped through the hose and torch assembly during welding cycles.

TRANSFORMER/RECTIFIER UNIT

The welding cycle of the power source is usually controlled remotely either at the welding gun/torch assembly by means of a trigger, or by a foot pedal

TIG WELDING

TIG TORCH

assembly. An optional feature is a time-sensitive contact switch which will close the gas and water valves at a pre-programmed time following a welding cycle.

## TIG Torches

The TIG torch is often referred to as a gun. Depending on how the unit is used, it can either be air or water-cooled. Water cooling adds greatly to the overall weight of both the hose and gun assembly so it is only a desirable feature if welding duty cycles and current capacities are high. The figure below shows a typical TIG torch assembly:

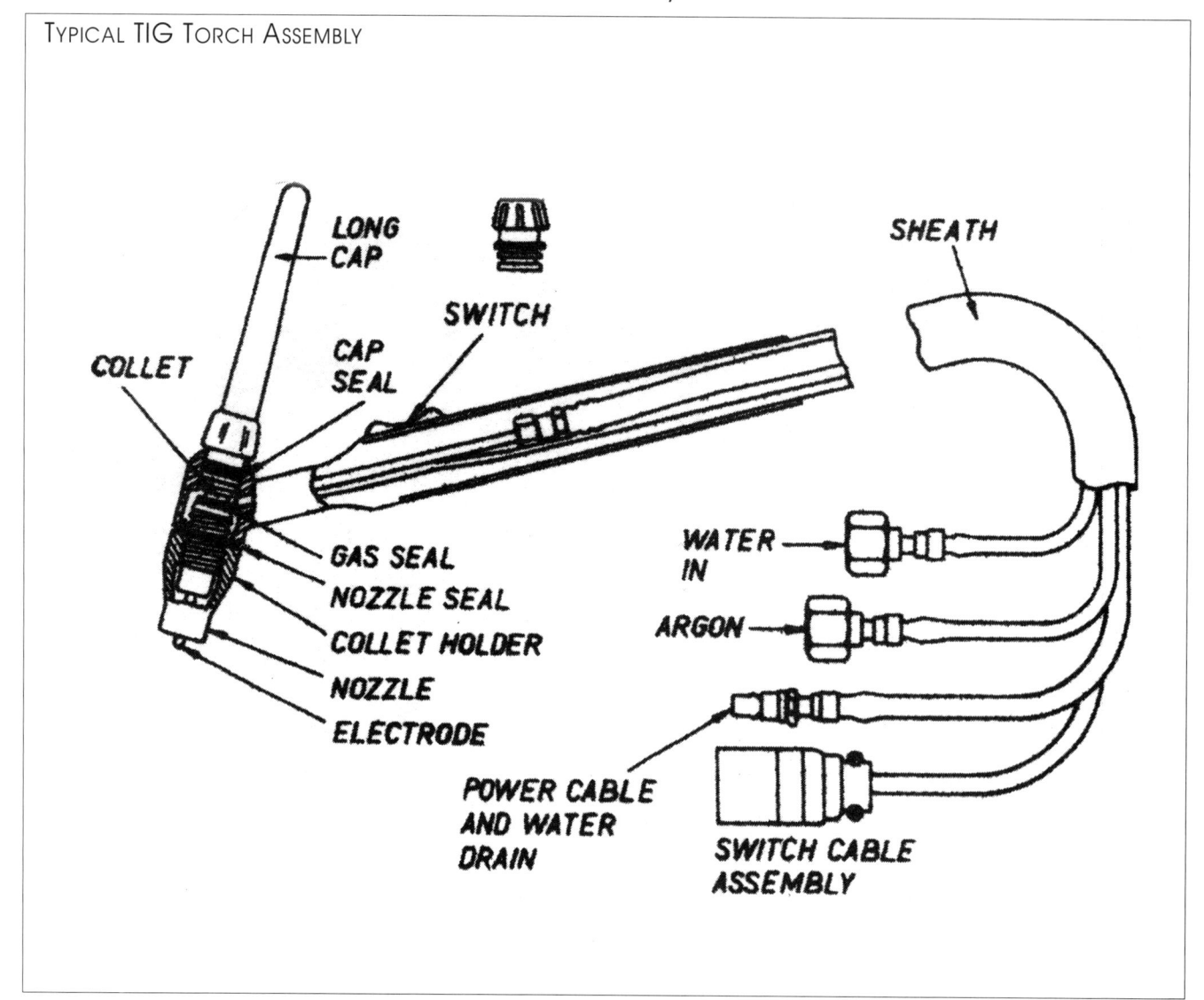

TYPICAL TIG TORCH ASSEMBLY

Labels: LONG CAP, SWITCH, SHEATH, COLLET, CAP SEAL, GAS SEAL, NOZZLE SEAL, COLLET HOLDER, NOZZLE, ELECTRODE, WATER IN, ARGON, POWER CABLE AND WATER DRAIN, SWITCH CABLE ASSEMBLY

70

## TIG Torch

The torch body holds a collet assembly which permits it to hold tungsten electrodes of a variety of diameters. The collet is used to lock an electrode into place and conduct electricity to the electrode. Because the tungsten electrode vaporizes slowly during welding, the welder is required to reposition the electrode from time to time to preserve the desired electrode protrusion from the nozzle.

The inert gas supplied to the torch is controlled by the remote trigger, either a torch trigger or foot pedal. The gas surrounds the collet and electrode and exits via the nozzle.

The nozzle is manufactured from alumina or silicon carbide ceramic. Both are highly fragile and will require periodic replacement. Heavy duty stations use water-cooled, metal nozzles which tend to last longer.

## TIG Electrodes

Electrodes are usually alloyed because their metallurgical properties influence the welds produced even though they are not integrated into the weld puddle (as would be in the case of a consumable electrode). Tungsten begins to melt at about 6107°F / 3375°C so only a little vaporization can occur during welding and in fact, the material should retain its hardness until contaminated with the weld material. TIG tungsten electrodes are ground with a coarse surface finish to allow the collet to grab hold of them when clamped in the torch.

There are no official AWS color codes used in categorizing TIG tungsten electrodes. However, the following color codes are widely used in the industry to designate tungsten electrodes:

Green: Pure Tungsten
Used for general purpose AC welding and can be used for DC welding.

Yellow: Tungsten Plus 1% Thorium
Used for easy start, good arc stabilization DC welding.

| | |
|---|---|
| Red | Red: Tungsten Plus 2% Thorium<br>Used for heavy current DC welding. |
| Brown | Brown: Tungsten Plus Zirconium<br>Used for easy start, high current AC welding operations where excellent appearance is a requirement.<br>Not recommended for DC welding. |
| Electrode Grinding | **Electrode Grinding**<br>Usually electrodes need to be ground to a not-so-sharp point prior to use. A sharp point is only required where the material to be welded is ultra-thin. Use a fine grit, high speed grinding wheel uncontaminated with previously ground materials such as aluminum and steel. When welding heavy materials with high current loads, grinding the electrode flat produces acceptable results.<br><br>During welding, if the tungsten electrode comes into contact with the material being welded it will immediately become contaminated. In effect, the tungsten 'alloys' with the base metal and this will contaminate the weld puddle. Cease welding, break the contaminated tip of the electrode away and grind to reshape the tip before reusing.<br><br>If the electrode tip produces a molten ball shape during welding without contacting the base metal, the cause is usually high current setting. Stop welding, dress the electrode and reduce the welding current. |
| Welding Gases<br>Aragon & Helium | **TIG Welding Gases**<br>TIG welding primarily makes use of argon and helium, and mixtures that use these gases in various concentrations. Both argon and helium are inert. A substance classified as inert means that it does not readily participate in chemical reactions and that is exactly the role of these gases in most TIG welding operations. Argon and mixtures that are primarily argon based are used for most manual TIG welding, while helium tends to be used for most automatic TIG welding. Helium is more expensive than argon and due to its lower density, 250% more is required to provide the same shielding protection argon achieves. |

Pure argon can be used as a general purpose shielding gas. More often it is blended to achieve specific results. Argon blended with 5% oxygen (sometimes known as argonox) is used to produce good appearance welds on aluminum. Unoxidized aluminum has a dull gray appearance. When an argon and oxygen mix is used, the aluminum surface of the weld bead is oxidized to produce a shiny finish.

Argon mixed with 5% hydrogen increases both welding speed and penetration when welding most stainless steels. Nitrogen gas can be used to weld copper providing it is de-oxidized.

## TIG Gas Regulators

The gas regulator functions pretty much like that used for MIG welding. It consists of a gauge and flowmeter. The gauge registers cylinder pressure. The flowmeter registers gas flow, usually reckoned in liters per hour. A flowmeter consists of an enclosed glass tube within which a plastic glass ball is floated: the higher the gas flow, the higher the float level of the ball in the tube.

REGULATORS

## Filler Wire

The filler wire used in the TIG process in most cases should be identical to the base metal to be welded. There are exceptions, for instance, the filler wire used to weld pure aluminum is usually alloyed with 5% silicon to improve the metallurgical properties of the weld puddle. Some stainless steels also require specially alloyed filler wire to ensure the welds produced are of at least the same strength as the original material.

FILLER WIRE

## TIG Current Characteristics

In the TIG welding process, the metal to be welded usually determines the current characteristic required of the power supply, unlike arc welding where the electrode flux coating is usually the critical factor.

CURRENT CHARACTERISTICS

The effect of the current characteristics used is shown in the figure on page 74.

# TIG Welding

## Effect of the Current characteristics Used

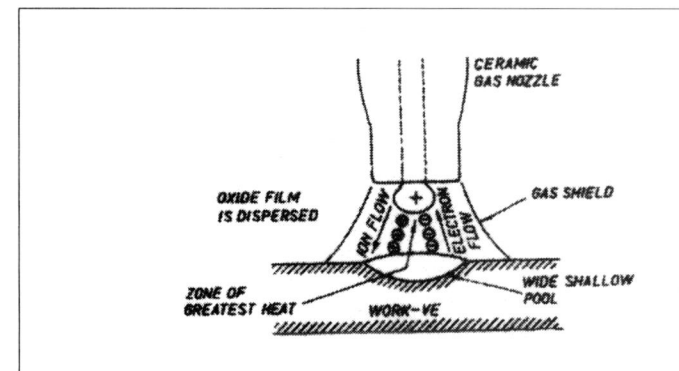

**A. DC TIG Electrode Positive:** More heat at electrode than on work. Used for welding ultra-thin material.

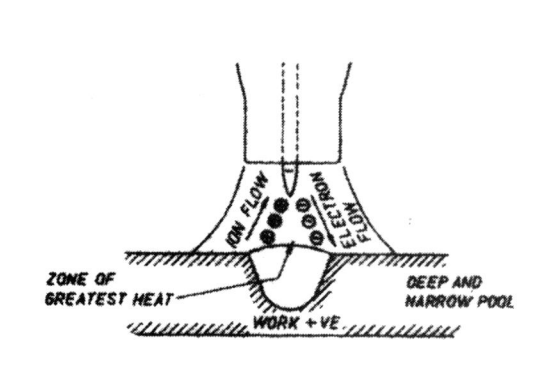

**B. DC TIG Electrode Negative:** More heat in weld puddle. Produces deep penetration into the base metal being welded.

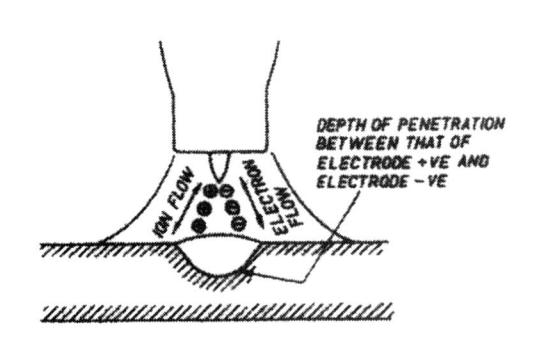

**C. AC TIG:** Produces a weld puddle characteristic mid-way between the extremes of DC electrode positive and negative characteristics.

## DC Electrode Positive

When DC electrode positive is used, heat produced is greater at the electrode than on the work, usually at a 65:35 ratio. This option is used for welding thin materials. It will usually mean that more attention has to be paid to dressing the tungsten electrode. The electrode should be ground to a fine point when welding ultra-thin material.

## DC Electrode Negative

The direction of electron flow is from the electrode to the work so the ratio of heat is reversed to 35:65. The result is that a higher percentage of heat is directed into the weld puddle with the result that:

➢ Base metal is heated faster.

➢ There is deeper penetration.

➢ Less frequent dressing of the electrode is required.

➢ Torch is less subject to overheating.

## AC Current

AC current has the ability to penetrate the refractory oxide coating produced on aluminum. However, because the arc extinguishes every half cycle, a high frequency current must be superimposed across the arc to maintain it. This superimposed arc provides a path for the AC arc to follow during the zero point in the AC cycle.

# TIG WELDING PRACTICE

Now you are ready to get going with some TIG welding projects. It takes little practice to produce some impressive results. Practice on relatively new, stock aluminum. This way you spend less time preparing and more time welding. You will need to set the TIG station to AC, select the recommended voltage and current settings for the thickness of your material. Regulate the shielding gas flow.

Use a stainless steel wire brush to clean the surface to be welded: don't skip this step, it will cost you many times over by forcing you to re-do the weld. Never use a regular steel wire brush or contaminated wire wheel because you will end up imparting more contamination into the weld area. Now try laying a few beads.

## TIG TECHNIQUE

*TIG Welding Technique*

Hold the TIG gun at an approximately 80 degree angle. Use the current control modulator to produce a weld puddle and dip the filler wire into the puddle. If you are right-handed, right dominant eye, the procedure is the same as you would use in oxy-acetylene welding. Hold the torch in your right hand the filler wire in your left. Start on the right side of your intended path of travel and move to your left. Dip the filler wire into the puddle at a 45 degree angle. In this way your dominant eye will be positioned to see exactly how you are progressing. And see, you will. More than any other welding process, you will get to look at exactly what is happening in the welding puddle, take corrective action where necessary, and yes, interact with the weld: this is one of the reasons you can produce such great-looking beads using this process!

# METAL BASICS 7

## METAL PROPERTIES AND IDENTIFICATION

It is necessary for a welder to be able to assess different types of metals. However, in order to determine the exact character of many alloys it is often necessary to contact the manufacturer. Perfectly executed welds that fracture soon after a repair have often been welded with an electrode that is not compatible with the alloy being welded. Or sometimes the problem is an improper understanding of the metal properties.

ALLOY COMPATIBILITY

Metals used in automobiles, trucks and trailers can be generally divided into 2 categories. These are:

➢ Ferrous metals

➢ Nonferrous metals

METAL CATEGORIES

Iron and its alloys are classified as ferrous metals. Nonferrous metals include metals and alloys that contain either no iron or insignificant amounts of iron. Some of the more commonly used nonferrous metals used on trucks and trailers are copper, brass, zinc, bronze, lead and aluminum. All metals can be welded once identified and the correct welding method selected. However, a large majority of the welding performed in a typical garage is done on ferrous metals ranging from mild steel, through low, medium and high carbon steels. Truck frame rails today are almost always fabricated from heat tempered steel that requires the use of special electrodes and welding techniques.

FERROUS AND NONFERROUS METALS

# Metal Basics

Of the nonferrous metals, aluminum is most commonly welded though seldom using a straight arc welding process. Aluminum and aluminum alloys may be welded using arc welding but better results are usually obtained using MIG and TIG processes.

*Steel and Iron*

## Steel and Iron
Steel is an alloy of iron. Iron is one of the more common 108± known elements on Earth. Most elements are naturally occurring, but a few are man-made.

Iron is used in its pure form in electrical/electronic components due to its ability to conduct magnetic lines of force. In electromagnetic fields, iron has low reluctance. Pure iron is also known as ferrite.

*Critical Iron Temperature*

## Critical Iron Temperatures
Cherry red - 1418°F / 770°C - iron becomes non-magnetic
Melts @ 2800°F - 1537°C
Boils  @ 4982°F - 2750°C

Steel is an alloy of iron (Fe) and carbon (C). A temperature of somewhere around 3000° F is required to alloy the two substances. As the carbon content increases in steel, the melting point tends to decrease. In steels, the carbon content seldom exceeds 1%. Small differences in the carbon content of a steel can radically change its character.

*Carbon Content of Steels*

## Carbon Content of Steels
Mild steel      0.1 - 0.4% Carbon
Middle alloy    0.5 - 0.7% Carbon
High alloy      0.8 - 1.0% Carbon
Cutting tools   0.7 - 1.0% Carbon

Few steels have more than — 1.2% Carbon
None have over — 1.8% Carbon

78

## Case Hardening

Case hardening is achieved by coating a steel surface with charcoal, limestone and tar, then heating to 1600°F / 871°C. This causes carbon atoms to migrate to the surface being treated. Quenching (rapid cooling in water or oil) will produce an extremely hard case and the shaft or bar core will remain fairly soft and resilient. Nitriding is another method of chemical surface hardening. All chemical hardening produces a hard skin, seldom more than 0.010" / .254 mm in depth.

Induction hardening is achieved by flowing a large electrical current through the metal to be hardened, followed by a controlled quench that tempers the outer surface. Induction hardening produces surface hardening to a much greater depth than chemical hardening processes—often as much as 0.060" /1.5 mm to 0.080" / 2 mm.

Annealing is a process that returns steel to its relaxed state. It removes any stresses but will also remove the effects of tempering. To anneal most steels, they should be heated to dull cherry red in the region of 1600° F and allowed to cool slowly. The blacksmith usually accomplishes this by covering the item to be annealed in insulating matter such as hot coal and ash.

When steels are worked at ambient temperatures, bending, tapping, turning, grinding, welding, hammering, stresses are set up. Annealing can relieve these stresses.

# ALLOYS USED TO IMPROVE THE METALLURGICAL CHARACTERISTICS OF STEEL

## Chromium

Small quantities improve the hardness and wear resistance. Large amounts (18%+) produce stainless steels (SS). SS can be 25 - 30% chromium.

METAL BASICS

COBALT
: **Cobalt**
Alloyed to increase hot hardness of steels used as cutting tools in HSS (high speed steels) used for machine tooling.

MANGANESE
: **Manganese**
Reduces brittleness and improves hardenability—that is, enables a steel to be tempered, especially those steels that have lower carbon content.

MOLYDENUM
: **Molybdenum**
In small amounts, improves deep hardenability properties of steel. Very commonly used in hand tools.

NICKEL
: **Nickel**
Improves toughness (fracture resistance). Frequently found in armour plating.

TUNGSTEN
: **Tungsten**
Increases wear resistance and hot hardness of steel; however it tends to increase the brittleness of steel.

VANADIUM
: **Vanadium**
Improves the forgeability of steel. It also increases harness and wear properties and has the effect of improving the toughness of most steels. It is commonly used as an alloy in the manufacture of hand tools.

CAST IRON
: **Cast Iron**
Much higher carbon content than materials classified as steels - between 1.5% and 5% Carbon with between 1% and 3% silicon added. Carbon in its form of graphite is used as the alloying agent. It is brittle and to some extent self lubricating so is often used as the material from which engine cylinder blocks and heads are manufactured. Cast irons can be alloyed in the same way steels are to produce characteristics not usually associated with cast iron such as fracture resistance and a high degree of flexibility.

# SAE CLASSIFICATIONS OF STEELS

4 Digit Numbering System

1st Digit
1. Carbon steel
2. Nickel steel
3. Nickel chromium steel
4. Molybdenum steel
5. Chromium steel
6. Chromium vanadium steel
7. Tungsten steel
8. No classification
9. Silicon manganese steel

2nd Digit
   Percentage of alloy

3rd and 4th Digits
   Percentage of carbon in 10/100

So, by using the above codes, an SAE 4140 steel would be a 1% molybdenum steel with 0.40% carbon.

# OTHER METALS AND METAL ALLOYS

Aluminum
Soft, relatively low tensile strength in pure form. When alloyed, characteristics alter and in fact, aluminum alloys may possess higher tensile strengths than mild steels. They are used extensively in the automotive and truck manufacturing industries due to light weight and good conductivity of electricity and heat. Although expensive, aluminum alloy is used as a truck and trailer frame material when light weight is required

Aluminum Copper
Structural components of chassis, buildings, aircraft and bridges.

Aluminum Boron
Conducts nearly as well as copper — much lighter — electrical wiring and terminals.

## Metal Basics

**Aluminum Manganese**

Aluminum Manganese
Good resistance to weathering and erosion. Used for highway signs, siding and roofing.

**Aluminum**

Aluminum
Aluminum is obtained from the ore bauxite. Alumina (aluminum oxide) is extracted from bauxite. The aluminum oxide is reduced to aluminum by electrolysis so it is often manufactured in locations with an abundant supply of cheap electricity such as Quebec. The alumina is dissolved in an electrolyte bath—electrical current is flowed through the solution—this action has the effect of separating the oxygen. The electrolytic cell is steel lined and faced with carbon; this serves as the cathode. Carbon electrodes are then suspended in the bath and act as anodes; this action releases $CO_2$. About 6 - 8 kw hours are required to produce 1 lb (0.5 kg) of pure Al. Canada produces no bauxite (this comes from locations such as Jamaica) but is the major producer of Al in the world due to cheap electricity.

**Melt & Boil Point**

Aluminum
➢ melts @ 1220°F / 660°C
➢ boils @ 4462°F / 2461°C

**Yield Strength**

Yield Strength
The maximum pressure a material will sustain before it permanently bends/twists out of shape. Measured in psi.

**Tensile Strength**

Tensile Strength
About 10% higher than the yield strength in most irons and steels, but it should be noted that the harder the steel, generally the less difference between the yield and tensile strength values. Tensile strength is the maximum pressure that a material will withstand before separating.

**Electrodes**

Electrodes
Are generally rated by the tensile strength of the wire.

## Metal Basics

**Hardened Truck Frame Rails**
Have a yield strength of 110,000 psi.

**A SAE Grade 5 Bolt**
Has a tensile strength of ±150,000 psi.

**Aluminum Truck Frame Rail**
Typically has a tensile strength of over 70,000 psi.

Aluminum is susceptible to oxidation/electrolytic degradation and deteriorates rapidly when exposed to road salt.

Aluminum oxide has a higher melting point than pure aluminum. This fact accounts for the difficulty in welding aluminum by non arc methods due to the relatively slow rate of heat rise.

Aluminum begins to oxidize almost immediately on exposure to air. The shielding gases used to weld aluminum in MIG and TIG processes often have a small percentage of oxygen added to provide a shiny, oxidized finish.

# WELDING TECHNIQUES    8

WELDING TECHNIQUE

Actual welding technique, that is, the laying down of beads, can be learned by most people in a relatively short time. There are exceptions. A small number of people, otherwise good technicians, are unable to weld, period. Try as they might, every time they touch welding equipment, a mess results. Usually a mess that takes a long time to rectify! If you are one of these people, the best strategy is to come to terms with the fact early on, and get someone else, anybody else, to weld for you. Comfort yourself with the knowledge that as an auto or truck technician, welding falls under the category of nice-to-know rather than need-to-know.

PREPARING MATERIALS

While simply running welds that look reasonably good can be learned with practice, the art of preparing materials to be welded requires a little more thought and the benefit of experience. The best welds can fail if incorrect methods, incorrect materials or flawed set-up procedures are used. It is important to know something about the:

➢ Metallurgical properties of the material to be welded.

➢ Welding equipment to be used.

➢ Filler wires or consumable electrodes to be used.

It is not uncommon to observe a perfectly executed weld that has failed. In such a case, though the technique of the welder cannot be faulted, the welder is nevertheless to blame. In other words, before attempting to weld anything, it is the welder's responsibility to determine what the material to be welded is, and to do some research on how the weld should be executed. This chapter will deal with some real world welding tasks. This is not intended to be a definitive guide. In most cases, there are a number of different methods that can be used to achieve the same end result. But most of the typical procedures outlined here do work and each welder should be able to adapt them to their individual style.

Caution!!
Some welding procedures are dangerous when performed on a vehicle. An automobile has miles of wiring conduit within highly flammable plastic coatings and a fuel tank. Flammable wiring coating becomes highly toxic when combusted and a vehicle fuel tank presents more of an explosion danger when empty than full. The work order may ask you to install a trailer hitch on a vehicle and say nothing about a leaking gasoline tank. Be observant before beginning to weld on any vehicle!

# OXY-ACETYLENE FLAME CUTTING EXERCISE

## Task
Perform a straight cut on a 1/4 inch mild steel coupon using an oxy-acetylene cutting torch. This exercise can be performed on the 6X3 inch coupons often used in welding instruction facilities.

## Equipment
Oxy-acetylene cutting equipment, goggles with at least a #5 lens filter, angle iron, clamps, soapstone, gloves, striker and tip cleaner.

## Mark Cutting Line
Use soapstone (ground to wedge shaped tip) and angle iron (or bar) to mark the cutting line on the steel coupon. Next clamp the angle iron so that when the

torch head is held against it, the center of the cutting tip is directly over the chalk line. In this way, the clamped angle iron will act as a guide to ensure a straight line.

## Set Working Pressures
Open the torch oxygen valve and set oxygen pressure to 20 psi / 138kPa, then close. In the same way, set the acetylene pressure at 3 psi / 21 kPa, then close.

Procedure

1. Open torch acetylene valve, ignite acetylene and set so that it burns smoke free. Then open torch oxygen valve and adjust so that the flame produced by the torch appears neutral. Press oxygen cutting lever to ensure that the preheat cutting flame remains neutral. Check that the area around the line of cut is clear.

2. Position torch vertically at the edge of the plate: the inner cone of the flame should be 1/8" (3.2 mm) from the metal. Heat beginning of cut to a cherry red color.

3. Press the oxygen cutting lever to begin cut. Angle the torch to about 10 degrees off vertical toward the direction of travel and move smoothly and evenly across the cut.

4. Evaluate the cut. The cut should have almost vertical cutting lines that are barely visible: there should be no melting at the top of the cut and no slag below.

5. Problems. Most problems are caused by either moving the cutting torch too quickly or too slowly across the work.

    Too fast: cutting lines (or grooves) curve in opposite direction to travel.

    Too slow: upper edge melted away, cutting lines are coarse and channeled.

    Unsteady travel: cutting lines that are erratic and uneven.

# GENERAL CHASSIS REPAIRS

**General Chassis Repairs**

Most of the welding performed by technicians on vehicles does not require a high level of skill. Autobody specialists are required to have a much higher level of welding expertise and this should be accompanied by having something of an artist's eye for the finished product. Before looking at some routine procedures found in auto, truck and heavy duty repair facilities, here are some simple rules to observe.

**Simple Welding Rules**

1. Use MIG welding whenever possible. It produces better results faster and requires lower levels of operator skills.

2. When you have to use an oxy-acetylene welding or cutting process, it takes MUCH longer to heat the material to melting temperature. Because of this, you should be aware of warping surrounding steel, destroying paint work, damaging wiring, destroying brake and fuel lines.

3. Disconnect the vehicle battery. In trucks and buses with isolation switches, open the switch and check with OEM service literature to determine whether any other precautions should be taken to protect the data bus and chassis modules.

4. When performing any kind of electric welding, locate the ground clamp as close to the weld area as possible. Electricity will always take the shortest route. It is possible to create arcing damage through lubricated engine and transmission components.

5. Have a fire extinguisher close to you at all times.

6. When performing any type of electric welding, ensure that the floor is dry around the weld area. Although welding voltages are comparatively low, high amperages are used and you increase the risk of electric shock when the work area is wet.

7. When using electric welding procedures, uncoil the both the electrode and ground wires used. Coiled

welding wires create an electromagnetic field that can amplify and distort the current characteristic you have selected. This can produce arc-blow. Arc blow is a characteristic in which the arc path from electrode to work is not direct, appearing to dance around the electrode. This can cause havoc with the weld puddle and severely affect penetration.

8. Some components on both automobiles and trucks are galvanized (zinc coated), for instance, exhaust system components. Zinc oxide fumes are highly toxic so great care should be taken to avoid inhaling them. MIG welding works well on galvanized steel but it does help if you grind away as much of the surface zinc as possible.

# WELDING VEHICLE FUEL TANKS

This is a commonly performed procedure that usually makes absolutely no sense. The exception is when a fuel tank has been damaged rather than corroded. In this case, evaluate the extent of damage and select the appropriate procedure outlined here. The method will depend on the fuel used in the tank and the extent of damage. When asked to weld a fuel tank by a customer, the smart thing to do is persuade the customer to replace the tank because, in the long term, this will usually be the cheaper option. Make sure it is possible to weld the tank: occasionally you will find tanks that have been internally corrosion-treated with substances that making welding impossible.

DAMAGED OR TREATED TANKS

## Gasoline Tanks

When I was an apprentice, I was told this story about technician X who would oxy-acetylene weld any fuel tank on the vehicle provided it was completely full of gasoline—it was probably embellished somewhat but of course this technician was reputed to have welded hundreds of tanks and evidently lived to tell the tale. This story may impress an eighteen-year-old, but hopefully no one who intends to make a living working

GASOLINE TANKS

on vehicles is stupid enough to attempt to replicate the method. A single metallurgical flaw in the tank or twitch of the torch arm and the result would be another statistic that does not get talked about much in service garages: the number of workers who die in the workplace. You might argue that such a death might merit a chuckle and a Darwin award, but the problem is that an exploding fuel tank would also endanger the lives of anyone else in the garage. If you have to weld a fuel tank, do it in a way that represents zero risk to you and your fellow workers.

### Steel Gasoline Tank Repair Method

*Steel Gasoline Tanks*

Gas tanks that fail due to corrosion from the inside out should not be welded. The cause is water that tends to sit in the bottom of the tank. The leak location is usually just the weakest point in a much larger corroded area, corroded to the extent it is almost impossible to weld. You can weld steel, but not rust.

*Safety Precautions*

If you elect to repair weld the tank take these safety precautions:

1. Drain fuel from the tank and store well in a sealed container away from the weld area.

2. Remove the fuel tank from the vehicle.

3. Remove the cap and all in/out fittings, sending unit and pump assembly from the tank.

4. Steam the tank for a minimum of 90 minutes. If the tank has internal baffles, change the position of the steam nozzle so that it discharges directly into each section of the tank. If the tank is of an irregular shape, double the steaming time.

5. Drain the tank (ensure there is no visible moisture in it) and allow it to cool. Plug the fuel tank cap and then discharge inert gas into the fuel tank, nitrogen is preferred.

6. If you are repairing damage, use MIG welding to first fill, then reinforce the damaged area. It is not a great idea to patch a tank because this type of

repair can cause problems later on, but if this is the method you have chosen, ensure that the patch is the same thickness as the tank material. Use a lap weld technique to weld the patch to the tank. If you are using an oxy-acetylene welding method, select a tip that allows for maximum weld penetration in the least time.

7. Clean out the tank interior with the steam washer. Air dry. Now apply spray type, die penetrant to the inside of the tank through the fill neck. Direct it at the weld area. Die penetrant is usually red in color. Next apply developer, usually white in color, to the outside of the weld area. The die penetrant will almost always locate any leakage point and a leak will show up as red in the developer. Remove the die penetrant, ensure that the fuel tank is clean and reinstall to the vehicle.

## Diesel Fuel Tanks

The appearance of a truck diesel fuel tank is usually more important than that on an automobile in which the tank is hidden. This should be considered when deciding on whether to repair or replace a fuel tank. Fuel tanks are often repair welded simply because a repair can get the vehicle on its way, rather than tie it up waiting for a replacement tank to arrive from Mexico.

Diesel fuel is also explosive given the right (wrong!) conditions and because of its residual oil content, more difficult to clean. Avoid attempting to weld any type of surface cladded (with chrome etc) tank. And what we said about corrosion failures of gasoline fuel tanks also applies to mild steel diesel fuel tanks so beware of getting into the business of attempting to weld rust. The good news is that most truck diesel fuel tanks are made from aluminum alloys so that welding them does not present difficulties.

WELDING TECHNIQUES

REMOVING CONTAMINATION

### Aluminum Diesel Fuel Tank Repair Procedure
The biggest challenge with effectively welding aluminum is removing contamination, especially the type of oil contamination that results from exposure to diesel fuel. Use a grinder and vee out the weld path. And if the weld has been caused by a crack, drill through at what appears to be the beginning and end of the crack.

REPAIR PROCEDURE

1. Drain fuel from the tank and store in a sealed container well away from the weld area.

2. Remove the fuel tank from the vehicle.

3. Remove the cap, vent, sending unit and all in/out fittings in the tank.

4. Steam the tank for a minimum of 90 minutes. If the tank has internal baffles, (most cylindrical tanks with a capacity exceeding 50 gallons do) change the position of the steam nozzle so that it discharges directly into each section of the tank.

5. Drain the tank (ensure there is no visible moisture in it) and allow it to cool. Plug the fuel tank cap and then discharge inert gas into the fuel tank: nitrogen is preferred.

6. I prefer to use TIG welding to repair aluminum tanks though MIG welding can also provide good results. Determine whether a patch will be required and remember to avoid using one if you can. Fill, then reinforce the damaged area using the TIG or MIG procedure. The filler wire should be a 5% silicon alloy: most aluminum fuel tanks are nearly pure aluminum and this makes a 5% silicon-cut filler wire ideal.

7. After welding, clean out the tank interior with the steam washer. Air dry. Now apply spray type, die penetrant to the inside of the tank through the fill neck. Stick your arm through the fill neck and direct the penetrant close to the weld area. Die penetrant is usually red in color. Next apply developer, usually white in color, to the outside of the tank around the weld area. The die penetrant will almost always

locate any leaks and a leak will show up as red in the developer. Remove the die penetrant, ensure that the fuel tank is clean and reinstall to the vehicle.

8. Reinstall the fuel tank to the chassis and plumb in the connections.

# WELDING A 6" PIPE IN POSITION

You might well ask what would be the point of an automotive or truck technician welding 6" pipe, but if you are serious about wanting to weld on vehicles, welding pipe in position will teach you how to tackle some of the trickier all-position welding. It is especially useful if you work on trucks or heavy equipment.

WELDING PIPE IN POSITION

Six-inch pipe has a standard wall thickness of .280" or just over 1/4 inch. If you butt weld pipe in-position, as you travel around the pipe you will work through every possible weld position from downhand, horizontal, overhead, and everything in between. You should learn how to do this using arc welding (the procedure described here), then graduate to MIG, which most people find easier.

## Method

1. Machine a 35 degree bevel in the abutting faces of the two sections of pipe to be joined. Finish by grinding a root face of approximately 3/32" (2.4 mm).

WELDING METHOD

2. Clamp both sections of pipe setting a gap of 3/32" (2.4 mm) between the two abutting faces. Use a 3/32" (2.4 mm) welding rod as spacers to set an even abutting gap. Next, use a level to ensure that the two pipe sections are aligned.

3. Tack, Select DC reverse polarity, set amperage at ± 90 amps and use an E-6010 1/8" (3.2 mm) electrode to burn in four tacks at 2, 4, 8 and 10 o'clock positions. Grind off excess tack weld. Check the alignment with a level.

## Welding Techniques

**Welding method**

4. Root Run
   Select DC reverse polarity, set amperage at ± 85 amps and use an E-6010 1/8" (3.2mm) electrode to perform a single pass root run. Begin at the 6 o'clock position and work around to TDC using a stroking action in line with the butt, ensuring that the action is one of burn through and fill. A 10 degree drag angle offset behind the direction of travel should be maintained. Repeat this action from the other side of the pipe joint, beginning at the 6 o'clock finishing at TDC. Grind off slag and any excess weld.

   Problems
   In the event of insufficient burn through, decrease the drag angle so that the electrode is at 90 degrees to the weld—that is, maintain a zero drag angle. If the problem is excessive burn through, increase the drag angle to as much as 20 to 30 degrees behind the direction of travel.

5. Fill Run
   Select DC reverse polarity, set amperage at ± 95 amps and use an E-7018 (low hydrogen), 3/32" (2.4 mm) electrode to fill the weld groove. This should be performed in a single pass on this size pipe. Begin at BDC and use a weave method with a zero drag angle to fill the beveled groove. Repeat the process on the other side of the pipe.

   Problems
   Slow travel or a long pause during weaving will cause the bead to droop. Too fast a weave will cause the center to sag forming a high crowned bead.

6. Cap Run
   Select DC reverse polarity, set amperage at ± 120 amps and use an E-7018 (low hydrogen), 1/8" (3.2 mm) electrode to cap the fill run. This should be performed using a wide weave that extends slightly beyond the beveled edge. Capping weaves are best performed using a zero drag angle. Again, perform two passes on either side of the pipe beginning each time at BDC and working to TDC.

# LENGTHENING A TRUCK FRAME RAIL

This procedure is common in new truck dealerships and if you are asked to perform it, you probably already possess good welding techniques, so the trick is in selecting the right method. Truck tractor frame rails are tempered. Tempering is a heat treatment process that in the case of a truck frame rail manufactured of a middle alloy steel, doubles the original yield strength of 55,000 psi, to around 110,000 psi. Applying heat to a hardened frame rail has an annealing effect that will soften the rail through the heat affected area. So it is important that the amount of heat applied during the cutting and welding is kept to a minimum. Truck frame rails are commonly C-channel formed with a radius between the flanges and web. Various sizes are used.

Truck frame rails are welded when a frame has to be lengthened. Because welding requires the application of a large amount of heat to the frame rail, the frame will be annealed surrounding the weld area. To minimize this annealing effect, the welding should be performed using a 60° or 45° weld angle: this will mean that the annealed area over any vertical section will be minimal. The method described here is field proven, but I want to emphasize that it is not the only method. As with most welding procedures, most of the work is in the preparation and set-up.

CAUTION!!
When you see a perfectly executed frame weld that has cracked dead center through the weld, the cause is almost always crystallization. Frame rails are middle alloy steels. Steel is an alloy of iron combined with a small amount of carbon. When the weld puddle becomes excessively heated, molten carbon tends cluster in crystallized pockets: this creates an extra-hard but brittle area that becomes susceptible to cracking. In the following procedure, the instruction is to allow the weld area to cool after each pass. Make sure you do. Impatience can cost you an expensive comeback repair!

*LENGTHENING FRAME RAIL*

*METHOD*

*FAILED REPAIRS*

WELDING TECHNIQUES

PROCEDURE

## Procedure

1. Getting Everything Level
It really pays to spend some time in getting the truck frame absolutely level before beginning. Get the drive bogies well out of the way. Assemble a collection of clamps, angle iron, and hydraulic jacks to support/adjust the height of the existing frame and the additional section.

2. Cutting the Frame
I prefer carbon arc to cut the frame. It is fast and neat and a bevel angle of around 30° can be produced on first cut. Cut at exactly 90° through the upper and lower flanges and at a 60° angle through the web. I like to bias the web cut so that it angles from top to bottom away from front, but many welders prefer the opposite. If you are using oxy-acetylene flame cutting, the process is slower and it is more difficult to establish the required bevel angle on a first cut, so it makes sense to grind the bevel after. Whichever cutting method you choose, use a right angle grinder to grind a butt face of around 3/32" / 2.4 mm. Prepare the C-channel frame section you are using to lengthen the frame in a similar manner.

3. Prepping for Welding
The butt gap should be set at 1/8" / 3.2 mm. Both the existing frame and the additional section must align perfectly. Use a level and squares to ensure the alignment is true. Then use an E6010 electrode at around 100 amps to burn in tacks, one on the upper flange, one on the lower flange, two in the web. The tacks should be high strength. Now, get out the level and square and check that the alignment is true. If it is not, grind out the tacks and begin again. It is rare to get this right first time so do not fall into the trap of thinking you can correct an alignment problem during the welding: it is guaranteed to get worse not better.

4. Root Run
Put in a root run using an E6010 electrode. My preference is to burn this in at high amperage

# Welding Techniques

(100+amps, reverse polarity), working quickly. Weld both upper and lower flanges using a vertical down method, use maximum penetration, sliding vertical-up motion through the web. Remember, at this point you are not concerned what the weld looks like. Allow the weld to cool to a hand touch. Check alignment once again. If your tack welds were sound, the alignment should not have changed, but check it anyway. Continue to leave all of the clamping and holding devices in place.

5. Fill Pass

    Grind the now cooled weld groove clean of slag and surface irregularities. Select a 1/8" / 3.2 mm. E11018 electrode and set the welder on reverse polarity at 120 amps. Run the weld as cool as possible so dropping the amperage setting down is OK. Use a downhand method to fill the flange grooves first: minimal weave with arc directed within the groove channel at all times. Drop the amperage down 5 amps and use vertical-up method to weld the web groove. Only one pass should be required to fill the weld groove. Allow the weld to cool to a hand touch.

6. Capping Pass

    Clean slag off the weld area. The capping run is used to relieve stresses that may have built up during the fill run. It should look good when you have finished but most of it will be ground away in the final step of the procedure. Use a 1/8" / 3.2 mm., E11018 electrode set at $\pm$ 110 amps on reverse polarity. A wide weave should be used that overlaps the dimension of the beveled grooved by the thickness of the electrode. Allow the weld to cool to ambient temperature.

7. Remove the Clamps and Supports

    Check alignment. This has to be perfect. Now grind away the cap weld and the protrusion of the root run. If you have done your job properly, there should be no evidence that any welding has taken place on the frame rail. Do not be tempted to leave the cap in place however pretty it may

PROCEDURE

## Procedure

appear: this changes the section modulus of the rail (a frame is a dynamic component, constantly oscillating) and will almost certainly generate a failure.

8. Conclusion

   There are no shortcuts when you are extending frame rails. The key to successfully welding on frames is in setting up the work properly. I know of technicians that use methods other than those outlined here that have produced good results so this method is not definitive.

Note: if welding non-hardened steel frame rails there is no need to angle the weld. Use a straight section groove and E7018 electrodes. Light-duty truck and trailer frames are usually manufactured from mild steel.

# Glossary of Techniques, Weld Faults and Welding Terminology

This section describes certain welding techniques and interprets the terminology. The techniques described sometimes apply only to one type of welding while others may apply to all types of welding process.

## Techniques

**Back-step Welding**  A welding technique in which welding beads are deposited in the opposite direction to the direction of progress of the weld run.

**Cap Run**  The final pass of a multiple pass weld joint, used to relieve stress and improve appearance. A weave technique is usually used in a capping pass.

**Fill Runs**  Term used to describe the passes laid on top of the root run in a multiple pass weld joint.

**Pass**  A single longitudinal weld run along a joint or weld deposit. The result of a pass is a weld bead.

**Root Run**  The first pass in a welding sequence where multiple passes are required. The term is generally used only with fillet and butt welds. The root run is often intended to provide maximum penetration and little or no side-to-side electrode movement is made. A root run is often performed with a different electrode than those used to fill the weld joint.

**String Bead**  A weld bead made without appreciable side-to-side movement: can be used to describe a root run.

| | |
|---|---|
| **Stringer Bead** | Same as a string bead or root pass, that is, usually the initial pass or bead. |
| **Tack Weld** | A weld (generally short) made to hold two pieces of metal to be welded into alignment until the final welds are made. Used for assembly and fabricating purposes only. |
| **Weaving** | A technique of depositing weld metal in which the electrode is rhythmically moved from side to side. Weaving tends to produce a relatively flat profile weld with low penetration but good appearance. It can remove some of the stresses produced by the fill runs in a welding joint. |
| **Whipping** | A term applied to the up-and-down sliding movement of the electrode used in vertical welding with some types of electrode to avoid undercut. |

## Weld Faults

| | |
|---|---|
| **Arc Blow** | Magnetic disturbance of the arc which causes it to waver erratically from its intended path. This is particularly noticeable when using direct current for welding in environments with other welding stations or in the proximity with other medium level radiation emission. Can be caused by welding with long lengths of cable coiled up: uncoiling the cable can remove the condition. |
| **Crystallization** | A condition caused when welding middle and high carbon alloy steels at excessive temperatures. Usually caused when welding multiple joint passes and insufficient cooling time is permitted between passes. The weld puddle becomes extra hard through the center. Failure is often identified as a technically perfect in appearance weld that fractures through the center of the weld. Crystallization is a real danger when welding hardened truck frame rails. For this reason, use no larger 1/8" / 3.2 mm electrodes and allow frame to completely cool between passes. Do NOT use MIG welding on frames. |

| | |
|---|---|
| **Gas Pockets** | A weld cavity caused by entrapped gas. Usually caused in multiple pass joints on middle and high alloy steels when insufficient cooling time is allowed between passes. A more severe form of porosity. |
| **Overlap** | Protrusion of weld metal beyond the bond at the toe of the weld. |
| **Porosity** | Gas pockets in metal caused either by overheating the electrode or the base metal during welding of steels. The condition may also be caused in cast iron welding by carbon clusters resulting from inconsistent casting concentrations. |
| **Slag Inclusion** | Non-metallic solid materials from electrode flux coating entrapped in weld metal or between weld metal and base metal. Usually the result of poor welding technique. |
| **Underbead Crack** | A crack in the weld joint zone not extending to the surface of the base metal. Often caused by crystallization resulting from overheating. |
| **Undercut** | A recess or groove melted into the base metal adjacent to the toe of the weld and left unfilled by weld deposit. Caused by high temperatures resulting from high current or voltage setting or poor technique. |

## General Welding Terminology

| | |
|---|---|
| **Arc Length** | The distance from the end of the electrode to the point where the arc makes contact with the work surface. |
| **Backing** | Non-fusible material (metal, asbestos, carbon, granulated flux, etc.) backing up the joint during welding to assist in obtaining a flush weld at the root. These may be strips, rings, welds, etc. Never used in pressure welding. |
| **Base Metal** | The metal to be welded or cut. |

# Glossary

| | |
|---|---|
| **BDC** | Bottom Dead Center |
| **Crater** | A depression at the end of a weld pass. |
| **Depth of Fusion** | The depth of fusion in a grooved weld joint is the area of the joint that has been molten during a weld pass. Varies with type of electrode used, polarity of DC current and welding technique. |
| **Ductile** | Pliable, not brittle. A steel that is ductile is capable of being drawn into wire. |
| **Filter Grades** | Eye protection filters are designed to remove harmful ultra violet light produced during welding processes. Filter grades #4 to #6 are designed for gas welding. Filter grades #9 to #14 should be used for arc welding. As the number increases, so does the protection level. Welders should select the highest filter grade number that enables them to properly see during welding. |
| **Flux** | The electrode coating or internal core, or gas used to shroud the arc and/ or weld puddle to prevent the formation of oxides, nitrides or other undesirable inclusions formed in welding. |
| **Heat Affected Zone** | That portion of the base metal which has not been melted but whose structural properties have been altered by the heat of welding or cutting. Especially significant when welding tempered or heat treated steels. |
| **High Carbon Steel** | Term used to describe steels containing 0.45% carbon or more. |
| **Low Carbon Steel (Mild steel)** | Steel containing 0.20% or less carbon. The most common steel used in vehicle construction typically rated at tensile strength values from 30,000 to 40,000 psi. |
| **Molten** | Matter in a liquid state. A weld puddle is in the molten state while it is in the process of being welded. |
| **Peening** | Mechanical hardening of metal surface area by means of percussive (hammer) blows or blasted shot. |

# Glossary

| | |
|---|---|
| **Penetration** | The distance the fusion zone extends below the surface of the components being welded. In a butt weld, the penetration of the root pass should cause a protruding bead to show on the reverse side of the weld. |
| **Preheating** | The heat applied to the work prior to welding or cutting. Necessary in some types of cast aluminum and cast iron to avoid cracking. |
| **Post Heating** | Heat applied to the work after welding or cutting. Necessary in some types of cast and cast iron to avoid cracking. |
| **Puddle** | The molten area in the base metal that includes molten filler wire produced by any welding process. Includes the weld bead and dilution zone. |
| **Slag** | Solidified flux residue that floats to the top of the weld puddle during welding, and protects the pass from oxidation during cool-down. |
| **Snorkel** | One of several types of welding fume extractor. Snorkels can be portable devices that can be wheeled up to a location in a shop where welding is being performed. Works like a big vacuum cleaner, pulling welding fumes and particulate in, filtering it, and discharging clean filtered air. |
| **Stress Relief** | The uniform heating of structures to a sufficient temperature to relieve residual stresses, which remain in a weld joint after welding, followed by uniform often controlled cooling. Stress relieving temperatures will vary with the type of metal used. |
| **Radiography** | The use of x-rays or gamma rays for the non-destructive examination of welded metals. Often used to observe critical welds in pressure vessels and pipe. |
| **Tensile Strength** | The maximum tensile (stretching force) stress required to cause a material to separate (usually expressed in pounds per square inch). |

| | |
|---|---|
| **TDC** | Top Dead Center |
| **Weldment** | Used to describe a welded assembly, the component parts of which are fused by a welding process. |
| **Weld Puddle** | That portion of a weld that is molten during welding. |
| **Yield Strength** | The maximum tensile (stretching force) stress required to cause a material to permanently deform (usually expressed in pounds per square inch). In non-hardened steels, this would generally be about 10% less than tensile strength. |

# Notes

# Notes